D0141990

SCIENTIFIC MODELS IN PHILOSOPHY OF SCIENCE

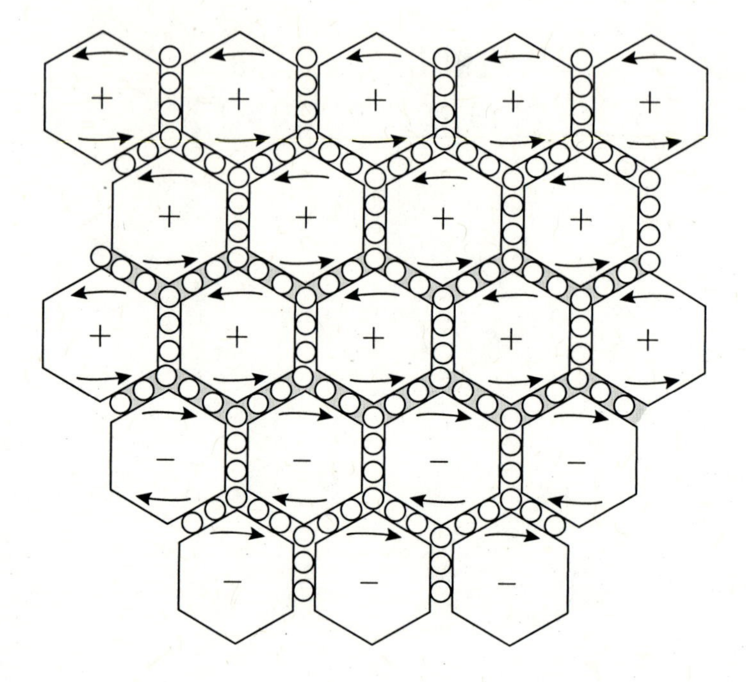

SCIENTIFIC MODELS IN PHILOSOPHY OF SCIENCE

Daniela M. Bailer-Jones

UNIVERSITY OF PITTSBURGH PRESS

Published by the University of Pittsburgh Press, Pittsburgh, Pa., 15260

Manufactured in the United States of America
Printed on acid-free paper
10 9 8 7 6 5 4 3 2 1

Library of Congress Cataloging-in-Publication Data

Bailer-Jones, Daniela.
Scientific models in philosophy of science / Daniela M. Bailer-Jones.
p. cm.
Includes bibliographical references and index.
ISBN-13: 978-0-8229-4376-1 (cloth : alk. paper)
ISBN-10: 0-8229-4376-X (cloth : alk. paper)
1. Science—Methodology. 2. Science—Philosophy. I. Title.
Q175.B163 2009
501'.1—dc22

2009013157

CONTENTS

FOREWORD

Daniela Bailer-Jones studied at the universities of Freiburg, Oxford, and Cambridge and from 1998 held positions in Paderborn, Bonn, and Pittsburgh, before setting up an Emmy-Noether research group in Heidelberg in 2005. She died on November 13, 2006, at the age of thirty-seven.

Daniela spent much of her working life thinking and writing about models in science and investigating how the concept of a model was used in philosophy of science. She was fascinated by the many different forms modeling could take and was deeply interested in their historical development and in how scientists themselves used and perceived models. She herself was no stranger to science, having earned a master's degree in astrophysics from the Cavendish Laboratory in Cambridge on the modeling of extragalactic radio sources.

This book grew out of Daniela's conviction of the central role that models play in science. In this endeavor she was at the forefront of a new movement in philosophy of science: a movement that took models, model building, and how models are used in science most seriously. Since the mid-1990s, the amount of attention paid by the philosophy of science community to models and their uses has increased considerably. It is an exciting new area of study.

As Daniela elaborates throughout this book, models were discussed in philosophy of science in earlier times, most notably in the late 1950s and 1960s by Rom Harré, Mary Hesse, and Ernest Nagel. But at that time the questions posed revolved around whether models could play a logical role in theory articulation and testing. The alternative was that they were just heuristic or somehow psychologically beneficial in nonlogical ways. In this manner they partook of the mysteries of metaphors and analogies in language, which were thought to be useful or important, but since they worked in nonliteral ways, they somehow had to be false descriptions.

The revived interest in models came when some philosophers, like Daniela, began to pay attention to the actual science and noted the almost ubiquitous role of models in many sciences. It became clear that many sciences had no "real" theories in the traditional philosophical sense. Models often played an important role in developing claims about the mechanisms responsible for producing phenomena and in testing hypotheses resulting from such developments. This book assays the history of discussions about models in philosophy of science, and makes some striking claims about how models really function in science. Daniela discusses how models relate to some traditional views of theories, how models model phenomena, and how models represent reality. These topics are being discussed in many contemporary circles in philosophy of science.

Daniela completed this book and submitted it for publication at the end of 2003, but it has taken several years for it to reach a wider audience. She received positive feedback on the manuscript, yet inevitably also some critical reviews, and had started to make some revisions. Since 1996 she had been undergoing cancer treatment. This became increasingly intense and exhausting, and at the same time she was also teaching, setting up a research group, undertaking other research, and being an active parent. When she died in 2006, the suggested revisions to the manuscript remained incomplete.

We decided that it would be too difficult to complete the revisions on Daniela's behalf and still keep the book in the spirit she intended. Given that she (as well as several reviewers) was happy with the original manuscript, we decided to publish the book as she intended. The only modifications are an updating of the bibliography (with thanks to Peter Machamer and Burlton Griffith for this undertaking) and minor language changes to the text. As the original publisher was not prepared to publish a posthumous work, we are grateful that the University of Pittsburgh Press, following the advice of its referees, agreed enthusiastically to take on this project. Although Daniela published various articles on different topics during her career, scientific models, it turned out, is her life's work. In this book we have her intellectual legacy.

Coryn Bailer-Jones
Max Planck Institute for Astronomy, Heidelberg

Peter Machamer
Department of History and Philosophy of Science
University of Pittsburgh

PREFACE

THIS BOOK TRACES the treatment of scientific models in philosophy of science. Taking my point of departure from nineteenth-century models in physics, I cover, in largely historical order, the topics of mechanism, analogy, theory, metaphor and paradigms, phenomena and representation. I elaborate what I take a scientific model to be and why I consider this notion of scientific model defensible. My goal is to show how models differ from theories and how they relate to the world. Part of this involves clarifying what phenomena are and what it means for a model to represent a phenomenon.

I review philosophical writings about models in philosophy of science, why models were for a long time disregarded as important elements of the scientific enterprise, and which important philosophical issues emerged when philosophers finally turned to models. So, to a considerable extent, my work belongs to the genre of history of philosophy of science. The more the historical treatment reaches recent times, however, the more the book becomes a guide to the burning issues concerning models as they are currently debated in philosophy of science: the model-theory distinction, phenomena, representation. Interestingly, the seeds for these debates can often be traced a long way back, but because the study of and the literature concerning models has often not been very systematic, many an insight concerning models was subsequently reburied. In response to this relatively troubled history of models in philosophy of science, I aim to give new impetus to the debate by making accessible relevant background information to the development of the philosophical treatment of models and by contributing my own thorough analysis of these issues.

I am very happy to acknowledge comments and suggestions on drafts of this book by a number of philosophers. Andreas Bartels accompanied my work on this book from conception to completion and read numerous drafts of chapters. Peter Machamer read and commented on the almost-final draft. I received valuable suggestions and pointers on individual chapters from James Bogen, George Gale, Michael Heidelberger, Wendy Parker, and members of the modeling reading group at the University of Pittsburgh in 2003.

Barrie Jones helped me in organizing the interviews with scientists at the Open University in Milton Keynes in 2001. I am most grateful to him and to all those who volunteered to be interviewed about scientific models and for their permission to be quoted.

I further acknowledge the financial support of a Emmy Noether grant of the German Research Council, 2001–2003, and a fellowship at the Center for Philosophy of Science of the University of Pittsburgh for 2003. The center and its fellows provided a wonderfully fertile environment in which to complete the writing of this book.

As always, I thank my husband, Coryn Bailer-Jones, for immeasurable support—intellectual as well as practical. I am also deeply grateful to my parents, Maria and Roland Bailer, for their care and support and for countless hours of babysitting. Katharina Rohlfing has been a most thoughtful friend to me, giving me inspiration and encouragement during all the years I have been working on this book.

I dedicate this work to Coryn and to Ezra, who keep me moving.

SCIENTIFIC MODELS IN PHILOSOPHY OF SCIENCE

SCIENTIFIC MODELS

1

IT IS MY AIM in this book to present views on models across philosophy of science. This is done with the view to highlight philosophical issues that arise when thinking about scientific modeling. To orient the reader, I begin with an outline of my position of what a scientific model is in section 1.1. In section 1.2, I present results from interviews with scientists from various disciplines to illustrate what scientists think scientific models are. In section 1.3, I briefly discuss methodological difficulties of a philosophical study of models, and in section 1.4, I give an outline of the whole book.

1.1 What I Take Scientific Models To Be

I consider the following as the core idea of what constitutes a scientific model: A model is an interpretative description of a phenomenon that facilitates access to that phenomenon. ("Phenomenon" refers to "things happening"—more on this in chapter 7.) This access can be perceptual as well as intellectual. If access is not perceptual, it is often facilitated by visualization, although this need not be the case. Interpretative descriptions may rely, for instance, on idealizations or simplifications or on analogies to interpretative descriptions of other phenomena. Facilitating access usually involves focusing on specific aspects of a phenomenon, sometimes deliberately disregarding others. As a result, models tend to be partial descriptions only. Models can range from being objects, such as a toy airplane, to being theoretical, abstract entities, such as the Standard Model of the structure of matter and its fundamental particles.

As regards the former, scale models facilitate looking at something by enlarging it (for example, a plastic model of a snowflake) or shrinking it (for example, a globe as a model of Earth). This can involve making explicit features that are not directly observable (for example, the structure of DNA or chemical elements contained in a star). The majority of scientific models are, however, a far cry from consisting of anything material, like the rods and balls of molecular models sometimes used for teaching; they are highly theoretical. They often rely on abstract ideas and concepts, frequently employing a mathematical formalism (as in the big bang model, for example), but always with the intention to provide access to aspects of a phenomenon that are considered to be essential. Bohr's model of the atom informs us about the configurations of the electrons and the nucleus in an atom, and the forces acting between them; or modeling the heart as a pump gives us a clue about how the heart functions. The means by which scientific models are expressed range from the concrete to the abstract: sketches, diagrams, ordinary text, graphs, and mathematical equations, to name just a few. All these forms of expression serve the purpose of providing intellectual access to the relevant ideas that the model describes. Providing access means giving information and interpreting it and expressing it efficiently to those who share in a specific intellectual pursuit. In this sense, scientific models are about empirical phenomena, whether these are how metals bend and break or how man has evolved.[1]

Models come in a variety of forms—that is, they employ different *external representational tools* that need not exclude each other. One form is that of ordinary language sentences written on a piece of paper. Take the following example of the hydrostatic equilibrium model of main sequence stars. Due to this equilibrium, such stars do not shrink or expand. According to the model, the gravitational pressure exerted by the very large interior mass of the star would pull the exterior layers of the star inward and would cause the star to collapse if this pressure was not resisted by the gas pressure (and/or radiation pressure) of the stellar material inside the star. The latter depends on the temperature inside the star. However, the star continuously loses energy via radiation from its surface. For the star not to shrink, this energy needs to be replenished, which is thought to happen through the process of nuclear fusion in the center of the star. Another form of communicating and highlighting aspects of this model would be mathematical equations, and yet another would be a diagram. Models can range from being objects to being theoretical, abstract entities, depending on the choice of external representational tools that represent the model. Examples of such tools include mathematical equations, plots of empirical data, sketches, objects, lists of assumptions, or simply ordinary language statements. These tools express information about theories employed in the model, empirical information about the subject matter of the model and various constraints arising from known and accepted laws of nature, as well as convenient assumptions that help to make the model more accessible and easy to communicate.

As scientific models are habitually contrasted with scientific theories, let me also comment on this distinction. In my use, theories are not about the empirical world in the same concrete sense as models. Models, by their very constitution, are applied to concrete empirical phenomena, whereas theories are not. Theories, in turn, have the *capacity* of being applied to empirical phenomena. This means that specific constraints belonging to a concrete case are inserted into the (intrinsically more abstract) theory (see chapter 6). A theory needs to be customized for its use in modeling a specific empirical phenomenon. This is what happens, for instance, in the hydrostatic equilibrium model of a star that uses quite basic equations from classical mechanics:

For a gas column of a given cross section, the hydrostatic equilibrium equation is $F_p \alpha \nabla P = -\rho g \alpha - F_g$, where F_p is the gas pressure force, F_g the gravitational force, ρ the density and g the acceleration due to gravity. For spherical symmetry, as in a star, we get $\frac{dP}{dr} = -\rho \frac{GM}{r^2}$, the equation of hydrostatic support, where M is the mass contained in a sphere of radius r and G the gravitational constant;

P, M, and ρ are all functions of r.[2] The model draws from a number of theories, such as hydrostatics and Newtonian gravity. Note that the way the equations are formulated already takes account of some of the constraints applying in a star—for example, that the pressure force, F_p, is equated with the gravitational force, F_g, and that a spherical shape is assumed. Here, constraints of the specific phenomenon to be modeled are already taken into account in the model. The equations, as I have quoted them here, are however still very general, because no empirical assumptions of the values of mass, radius, density, and so on have been inserted and used to test and confirm the model.

Due to the inserted constraints (for example, the shape of a star, its mass density) and other elements of knowledge about the phenomenon to be modeled, a model can be molded and adjusted to address the real and concrete empirically observed situation. Consequently, a model is never as general as a theory. For various reasons, it is difficult to be so specific as to match the empirical phenomenon perfectly, but in contrast to theories which *aim* to be general, models *aim* to match quite specific empirical situations well.[3] While theories tend to aim at addressing the empirical world through abstract principles and are expected to be globally valid, models apply locally (indeed, they often are a way of applying theory locally), and when models contain theory elements, they adjust them to the concrete empirical situation to be modeled. This appears to be the dominant way in which such theory elements can be tested empirically. Only for a model can it be decided concretely whether and to what extent the model applies to the empirical world.

According to my explication, more or less anything that is used in science to describe empirical phenomena is a model, and indeed this seems to be the case. The tools employed to grant us and others intellectual access are of great diversity, yet this in itself is no reason to deny them the status of being constituents of models. Modeling in science is pervasive; it has become increasingly varied and more and more theoretical. The sheer diversity of models makes it unlikely that a narrow account for *all* models can be given, so my preliminary account and example need to suffice as a point of departure.

1.2 What Scientists Take Scientific Models To Be

It is my intention to develop a concept of scientific models that is in agreement with the way this concept is used by scientists. For this reason, I asked scientists personally about their views on models by way of qualitative interviews (Bailer-Jones 2002). As the name suggests, a qualitative interview is not one where the results are made subject to measurement and quantification. Instead, the interviewee is treated as an expert on the matter at issue and given space to elaborate his or her views. On the other hand, it can then be expected that the interview reflects the interviewee's views much more truthfully and in depth than in the case of a quantitative interview, in which the interviewee would be restrained by a narrow canon of questions and restricted in his or her range of reply. The interviews were recorded and fully transcribed. Methodological details and considerations are described in ibid.

In accordance with the technique of a qualitative interview, I started off the interview with some words of introduction to the topic but avoided reference to existing philosophical work on scientific models and prevailing styles of acquiring views on models. I emphasized that my intention was to find out about scientific practice in various disciplines and about what scientists themselves think about models. I had prepared a set of questions, beginning with the scientists' discipline and subject then asking them whether they used models and, if so, to give me their personal view on what models are. Then I asked for the kind of models commonly used by the interviewee and what one gets out of their use. Toward the end of each interview, I asked about the relationship of models to empirical reality. As it turned out, I was rarely able to ask the majority of my prepared questions, because they were generally addressed in the natural course of the conversation. The scientists did not require additional prompting to bring up the major issues I wanted to cover. The technique I adopted was to listen to subjects attentively, to encourage them to expand and develop their ideas. Sometimes I picked up a formulation introduced by the scientist and probed him or her further on its meaning and consequences, or I would suggest consequences or further developments to proposals an interviewee had made in order to explore whether he or she had really meant what he or she seemed to have said. Most of my additional questions were directed toward avoiding my misunderstanding or further clarification of points raised by the interviewee.

The interviews lasted between thirty-five and seventy minutes. The interviewees ranged in age from the late twenties to the late sixties, with Andrew Conway being the youngest and Colin Russell being the oldest, and the majority of interviewees being in their fifties. The older scientists have considerable experience in teaching their subject and also have experience in research, even if some were no longer actively doing research. Only one of the interviewed scientists was female, Nancy Dise, a biogeochemist. In sum, although these interviews may not be representative for all of science, because the sample of interviewed scientists was small and subject to various selection effects, the statements made are nonetheless illustrative of views on models that prevail among scientists. The fact that the views expressed were not diametrically opposed to each other, despite the range of personalities and disciplines involved, is reason for some confidence. Of course, different scientists emphasize different points, depending on their discipline, preferences, and worldview, but it was certainly not the case that *widely* diverging views were expressed. Let me now present some of the interview findings. The quotes given here do not replace a full analysis, but they are representative snapshots giving a taste of the positions taken and the ideas put forward.

I begin with the views given by scientists when asked what scientific models were. I asked this in the early stages of every interview and again at the end. Not everybody attempted to capture the meaning of a model in such a definition; some denied it was even possible.

> [A] model is the way you set up a problem for investigation. . . . Actually, when someone says "scientific model" to me, you have to get away from thinking of it as just a geometrical model or a visual model, because there are many other sorts of models. You can actually have a model as a set of statements.
>
> *Barrie Jones, physicist and planetary astronomer*

[M]y understanding of a model is a formal statement of a hypothesis designed to be tested.

Peter Skelton, evolutionary palaeontologist

I think a model is a way of summarizing all important fundamentals that go into describing a natural phenomenon, that you have obtained observations of. And it contains a number of assumptions stating what you want to be concerned with in modeling, and requires you to input data from the real world that will predict the observations you have designed it to predict.

Andrew Conway, solar physicist

So the real world does what it does and the model is some simplified representation of part of it in one form or another, and in my case generally in the form of equations.

Robert Lambourne, particle physicist

[I]t will be mathematical; it will have elements within it which have some correspondence with the elements of reality, and the relationship between the elements in your model should be such that when you put it together and allow . . . the consequences of the mathematical structure to follow through, it should have an ongoing correspondence with a large number of facts which correspond to experiments, which weren't immediately within the fact which you first put down as the elements of the model.

Ray Mackintosh, nuclear physicist

[A] model in science generally is a means of depicting on paper or in terms of hardware what is something at the molecular reality. . . . Any representation of what is going on at the molecular scale by definition must be a model, because it can't be the real thing. Any representation whatsoever, be it of a simple structural formula, a steric formula, a confirmational formula, . . . a formula of showing all kinds of things getting together and rearranging and reacting, all of that, is a model.

Colin Russell, chemist and historian of chemistry

There are a number of characteristics of models that were identified by interviewees as being typical of models. I list these here and illustrate them with exemplary quotations.

The formulation of a model depends on *simplification* and the use of *approximations.*

> Generally I would consider a model a simplification of the system, incorporating what you consider to be the most important elements of that system. So you are describing the system, but you are not describing it in all of its detail.
>
> *Nancy Dise, biogeochemist*

> I tend to reserve the word [model] for something that is not the best theory we have of something, but for which you are deliberately making approximations.
>
> *John Bolton, solid state physicist*

By simplification it is attempted to *capture the essence* of something. This essence is often pursued at the cost of *omitting other elements and details* that might be considered in the model. This can result in limited validity of the models.

> So I see a model as really trying . . . to identify the essence of the observations that you are trying to explain, as simply as possible.
>
> *Andrew Conway, solar physicist*

> [T]he scientific model is something which captures enough of that reality to give maybe not a quantitative description of reality, but at least a qualitative description of what's going on. But in order to achieve this it has to throw away some parts. . . . But a well-designed model will encapsulate the essence.
>
> *John Bolton, solid state physicist*

> What determines an exclusion is either ignorance or a deliberate deselection as being irrelevant or problematic or simply impractical.
>
> *Colin Russell, chemist and historian of chemistry*

> So, what we now see, the kind of models which I apply are understood to have restricted validity within nuclear physics as a whole because there are certain regimes where those models do not apply.
>
> So there are many things which are left out and, of course, one could say in a holistic calculation you got to put them all in. But you can get some indication of the validity of those approximations by putting particular things in one at a time.
>
> *Ray Mackintosh, nuclear physicist*

When models leave out certain details and as a result are of limited validity, then *multiple models* of the same thing can occur.

> [Y]ou'll need different models to describe different things and different aspects of reality.
>
> *John Bolton, solid state physicist*

> And indeed, nuclear physicists themselves recognize a hierarchy of these models. They recognize how some are related to others, and so on. So they are not under the illusion that all those models are equally valid and equally useful. They have different functions, different purposes, different limitations.
>
> *Robert Lambourne, particle physicist*

Even if models are simplified, even though they focus on some details and omit others, they nonetheless need to meet certain objective criteria. Models have to maintain a *link to the empirical data available* on what is modeled.

> [M]odels clearly must have some relationship to empirical data or they wouldn't be models.
>
> *Colin Russell, chemist and historian of chemistry*

> I wouldn't be comparing my models precisely with the features in the data. I'd be comparing very broad brush.
>
> *Andrew Conway, solar physicist*

> [A model may not hold up to subsequent observations,] but for the information you have currently available, it's the best model around.
>
> *Peter Skelton, evolutionary palaeontologist*

Because of this expected link to empirical data, a model can be subject to *empirical testing* and it can give rise to *predictions*. Successful predictions act as a sign of confirmation of a model.

> From my own personal perspective, and no more than that—I'm not going to say that I'm right or wrong, I'm simply saying it's my taste—I prefer to be able . . . to test whether [a model] is right or not.
>
> *Malcolm Longair, astrophysicist*

> For me, a model is effectively a formal statement of a hypothesis which is designed to be tested. A hypothesis may be a generalized idea about the operation of some process or some historical pattern, but in order to test that you need to have a precise statement of what that process consists of and what its effects are likely to be in given circumstances. And I suppose I would see a model as being that formal statement which is the step prior to testing. The function of a model is to be tested.
>
> *Peter Skelton, evolutionary palaeontologist*

> And a model is good insofar as it provides results which agree with experiment. . . . Models are successful if these predictions turn out to be correct.
>
> *Robert Lambourne, particle physicist*

> You allow the mathematics to work itself out and this should provide an ongoing correspondence of a wider range of contacts with reality than were within it at the time when you set it down.
>
> *Ray Mackintosh, nuclear physicist*

Although agreement with empirical data is expected, the model needs not necessarily be correct.

You've got some plausible models which comforts you because you can think, well, this is not a total mystery to me; I can imagine what might be going on here. I don't actually know what's going on here, but this is all right, we got some ideas.

Barrie Jones, physicist and planetary astronomer

If it is possible that the model is not right in every respect, it must have some other benefits that compensate for this deficit. Providing *understanding and insight* is such a positive outcome that is frequently expected of scientific models.

So sometimes getting, I suppose, a possible match to reality is not everything. What you are looking for is an understanding of what's happening in nature, and sometimes a simple model can give you that, whereas a very large computer program can't.

John Bolton, solid state physicist

[Models] act as a medium for thinking about scientific problems. [A model is] an aid to understanding and an aid to thought.

Barrie Jones, physicist and planetary astronomer

[Y]ou get new insight at different levels into the ways in which the nature is working.

Malcolm Longair, astrophysicist

Besides highlighting the characteristics of models, there is a further, more general issue of how central models are for doing science and how important they are to scientists.

I think the word "modeling" is much more used now amongst theorists, amongst physicists, amongst mathematicians, than it used to be. So I think there is a much greater consciousness of modeling than was the case when I first started in physics, and certainly in generations before.

Robert Lambourne, particle physicist

I feel the most exciting thing in science is building the model, the idea of building it, rather than taking somebody else's computer code, say, and then adapting it. This doesn't really interest me on a personal level, it doesn't stimulate me to do that.

Andrew Conway, solar physicist

I spend my life doing models.

Ray Mackintosh, nuclear physicist

I guess, somewhat reluctantly, I would take the view that physicists are almost entirely concerned with the process of modeling, that what they do, certainly in theoretical physics, is to try and form mathematical models that represent large or small portions of the universe, common or uncommon processes.

Robert Lambourne, particle physicist

I would clearly underline that essentially the whole of physics, whether you like it or not, is actually building models all the time.

Malcolm Longair, astrophysicist

Contemplating the role of models in science, however, also brings up the issue of how models relate to theories. This distinction is not clear-cut, but there exist strong intuitions on the matter.

I'm prepared to use the word "theory" for the most accurate descriptions we have of reality which we may even believe are true.

To me a theory makes grander claims than a model. The theory claims to be the way things work. And if they don't work that way, the theory is wrong.

John Bolton, solid state physicist

I think theory is . . . somehow at a higher level than a model, I would think. It's a more meta-level. The model is kind of below. It's some kind of reflection of theory. It can incorporate theory.

[T]o construct a model you can draw on a theory and apply it to specific circumstances.

Barrie Jones, physicist and planetary astronomer

Ultimately in physics, [fundamental or universal rules are, for example,] Maxwell's equations, fluid equations, cross sections that come from particle physics, all, I mean, fundamental, physical laws of nature.

[Fundamentals are] what you can't argue with.

[T]here is nothing more fundamental that [these fundamental rules] are understood in terms of.

[W]hat a good model will do is select a small plausible subset of all these fundamentals and give you insight into how that subset conspires to produce the phenomena you have observed.

Andrew Conway, solar physicist

There is no, in my mind, no real linguistic significance to the distinction between theory and model.

Robert Lambrurne, particle physicist

Let me summarize some broad conclusions from these quotations:

- Models are acknowledged to be central to today's scientific enterprise.
- Theories tend to be seen as more fundamental and more general than models, while models can draw from theories and thus show how theories contribute to modeling specific phenomena.
- There is a sense that models provide insight and contribute to our understanding of the natural world.
- Models simplify things and thereby try to capture the essence of something, while they leave out less essential details about the phenomena modeled.
- Models can be of limited validity, which means that different models can fulfill different functions. So models can be useful without it being known that they are "right," or even when they are known not to be right.
- Models should not only match available empirical data, but they should also give rise to predictions and that way become testable.

These opinions that were expressed give a foretaste of many of the philosophical positions to be retraced in the chapters that follow.

1.3 How to Analyze Models

The first two sections of this chapter are in some ways representative of a deep-seated dilemma: How are scientific models to be studied? There is what philosophers think, and there is what scientists think, *and* there is what scientists practice. In fact, often philosophers examine how models are used in scientific practice. Correspondingly, at least two groups contribute to the debate on scientific models. One group is scientists, because they are the ones who use scientific models. I have already introduced one method by which to study their views: conducting interviews. The other group is philosophers of science who conceptualize scientific models. They can either take a theoretical approach or approach the issue from what they find in scientific practice. Considering and criticizing philosophers' views on scientific models will take up most of the rest of this book.

Scientists can apparently afford to take a wholly pragmatic approach to models.[4] They take the liberty to call an enormous range of things models, simply depending on their discipline, their preference, or some historical quirk. There is the big bang model versus the steady state model in cosmology, the Standard Model of elementary particles, Bohr's model of the atom, the DNA model, models of chemical bonds, the billiard ball model of a gas, or the computational model of the mind, to name just a few. Corresponding to this range, scientists' conception of "model" has been prone to change and to develop over the years and across research traditions. Scientists frequently choose freely among the labels "model," "theory," or "hypothesis," and the choice is for them mostly of very little consequence. The issue of importance for them is that this "something" serves the purpose and the expectations for which it has been designed. Anything else can, in keeping with a pragmatic approach, be perceived as a side issue, from their perspective.

As a philosopher of science, when I address the topic of scientific models and query the concept of a model, then definitions, stipulations, and conceptual clarifications are required because they are of consequence philosophically. For instance, there is a tradition of contrasting models with theories. Consequently, it is necessary to locate new work on models in the tradition of philosophizing about models and theories and to specify this relationship. Any new formulation of a philosophical position is partly a response to the failings of its forerunners. If possible, such a new formulation may be specifically geared toward addressing points that appear unsatisfactory in earlier positions. Any new formulation of a position is likely to depend quite heavily on an understanding or an awareness of its forerunners (and their perceived failings). This not only holds for my position on models that I present in this book, but it is a general requirement for all those who concern themselves with any philosophical position. Therefore, to promote an informed discussion about scientific models, I dedicate a large part of this book to giving a comprehensive account of the philosophical stances taken toward models.

One important lesson from the history of philosophy of science is that there is more to the analysis of science than rational reconstruction and retrospective justification of scientific knowledge. The study of scientific practice has revealed some interesting and unexpected views of how empirical data are gathered, how phenomena are interpreted, and how theoretical conclusions are drawn—in short, how science progresses and develops. Crucially, for my interests, studying the scientific practice uncovers the importance of models in this process of science. However, studying the scientific practice also means, with regard to the current topic, taking on board how scientists use models and how they employ the term "model." Thus scientific practice poses constraints on a viable philosophical concept of scientific models. Paying attention to scientists' sometimes quite free uses of the term, the task of philosophers is then to avoid the trap of calling everything a model and to preserve the concept of a model as a useful and meaningful term for philosophy of science.

It is therefore a worthwhile challenge to come up with a conception of models—and of theories—that does not ignore (and attempt to override) the diversity of model uses; diverse uses of "model" need to be accommodated somewhere in the system-to-be, even if this results in a fairly wide conception of "model." So one aim is to provide conceptions of "model" and "theory" that are reconcilable with the ordinary understanding of these terms by scientists. The other aim is, however, that the proposed framework for the philosophical consideration of models and theories is systematic and succinct enough (that is, not "drowning" in the diversity encountered in practice) that philosophical issues can be posed, considered, and reflected upon. Note that if too loose a use of "model" were permitted, nothing beyond the use of the term in an individual case could be known about models, and it would not be possible to treat models as a more general subject of investigation. In contrast, my vision for such a proposal of a revised and clarified philosophical concept of model is that it offers opportunities for novel perspectives and alternative solutions to such issues as, for example, scientific realism that have become overfamiliar in the more traditional context of treating theories and models in philosophy of science. Explorations of such perspectives and solutions can be found throughout this book.

1.4 Outline of the Book

I see it as my task throughout the book to strengthen the outlined account of models. Before I focus on this, however, I shall guide the reader through the thinking on scientific models. This is done partly historically and partly thematically. The chapters roughly follow a historical order, beginning in the early nineteenth century and concluding with the issues that occupy us at the turn of the twenty-first century. As each chapter also traces a historical development with regard to the treatment of a certain topic, there is sometimes, from the end of one chapter to the beginning of the next one, a considerable jump back in time. This is so because I include some recent views on the issue at stake at the end of most chapters. Correspondingly, some chapters overlap with others in the stretches of time they cover, yet the overall direction remains from past to present.

Models had a poor reputation in the first half of the twentieth century. They were considered as an inferior tool used by those who could not do science "properly." The first line of argument is intended to explore the historical circumstances that led to the past bias toward theories and against models. These historical circumstances have to do with (1) the use of models in nineteenth-century physics, and (2) the tradition of Logical Empiricism in philosophy of science. The first of the reasons why models have historically been disregarded is the hostile attitude that (some) scientists, or physicist-philosophers, developed at the end of the nineteenth century toward models. This was at a time when physics became increasingly abstract and mechanical models consequently appeared less and less appropriate. The use of models in the nineteenth century and the attitudes toward models and theories that developed at that time are examined in chapter 2. The appeal of mechanisms, explicit in the use of mechanical models and implicit in many modern models, is a further topic of that chapter.

In chapter 3, I pick up a thread that also goes back well into the nineteenth century—namely, the role of analogy. In fact, some of the constructs that were referred to as analogy in the nineteenth century we would most likely call models today, only that models in the nineteenth century were pretty much exclusively physically tangible models. Analogy had been established as an important instrument and strategy in science by such scientists as John Herschel and James Clerk Maxwell. Their views on analogy are presented, followed by philosophers' and psychologists' treatment of the topic in the twentieth century. Although analogy is not synonymous to model, it is often an important element in scientific modeling. Analogy is not only responsible for a great deal of the heuristic advantages attributed to models, it has also been instrumental in overcoming the limitations of thinking of models purely as mechanical in the narrow sense of the word. Analogy helped to lift our understanding of what a model is, and of what counts as a mechanism, to a more abstract level. Drawing an analogy often requires a process of presenting something appropriately—that is, at a more abstract level—to make the analogy applicable. For example, if both light and sound display wave properties, then these wave properties must be expressed in a way not exclusively specific to either light or sound; the wave equation then is a "re-representation" of the wave properties and is more abstract than the original descriptions of light and sound waves.

The theory-biased perspective of scientists is reflected in the views of the early scientific philosophers of the Vienna Circle, explored in chapter 4. These philosophers formed the movement of Logical Empiricism that remained influential in philosophy of science for many decades thereafter. Issues such as scientific discovery and theory change were particularly hard to accommodate in a Logical Empiricist frame of mind, such that only in the 1950s did a small niche develop in which consideration started to be given to scientific models.

Chapter 4 is dedicated to answering the question of why scientific models were essentially not discussed and considered—and if at all, then negatively—in the first half of the twentieth century.

The turn toward models, which began in the early 1950s, intensified in the 1960s. At that time a number of philosophers, first and foremost Mary Hesse, focused on reasons in favor of the study of scientific models. Such reasons were to explain discovery and creativity. With Thomas Kuhn's attention to scientific revolutions, the whole issue of conceptual change entered the philosophical debate. Chapter 5 deals with accounts of conceptual change, Kuhn's paradigms, and Hesse's metaphor approach to scientific models. Although models continued to be viewed as merely an addition to scientific theories by many philosophers who saw them as if not dispensable then at least as subordinate to theories, a second approach to the study of models flourished in the 1960s.

This approach is much more than the metaphor approach in the formalist tradition of the Logical Empiricist philosophy of science. This work adapts tools from mathematical model-theory for the philosophical treatment of models. Commencing with the work of Patrick Suppes in the 1960s, it later developed into the so-called Semantic View of theories. I describe these formalist approaches in chapter 6 and contrast them with those that explore scientific models by studying the scientific practice through case studies. Nancy Cartwright's work has been influential in this context. The case study approach assembles further evidence about the centrality of models for doing science, resulting in the claim that models are mediators between theories and the world (Morgan and Morrison 1999). The tension between the Semantic View of theories and case study–based approaches provides fertile ground for proposing a distinction between models and theories at the end of the chapter.

Answering the question of what models are about by saying that models are models of phenomena requires some attention to empirical aspects of what it is that is modeled—that is, what phenomena are. This is the topic of chapter 7. One lesson to learn from the so-called new experimentalism is that observation itself is rather a vague term. Data need to be statistically analyzed and cannot be equated with the phenomenon that is supposedly observed (Bogen and Woodward 1988). My own approach, which builds on the model-theory distinction of chapter 6, is to argue that the connection between theory and observational data is built into the modeling process. However, this attributes a role to scientific theories in which theories are not something that refers to empirical phenomena—or rather, only via models. In models, theories are customized with respect to a specific phenomenon.

Another important topic to reconsider in the light of a revised understanding of scientific models is scientific realism. Some current work claims that models *represent* phenomena of the empirical world. This raises the question of *how* scientific models represent. I deal with this topic in chapter 8. Chapter 9 provides outlook on the wider philosophical implications of the study of scientific models. Recapitulating the topics and findings of chapters 2 through 8, in this final chapter our modes of access to the understanding of scientific models are reconsidered and evaluated. Having argued throughout the book that the potential of scientific models as a working conception in philosophy of science has been, if anything, underestimated, chapter 9 takes stock and looks ahead to what further progress may be made in dealing with a range of philosophical issues emerging from the study of scientific models.

As this book shows, for a host of reasons philosophizing about models has sometimes been not much more than a waste product of philosophizing about theories. This situation resulted in a considerable neglect of the topic during the one-hundred-year history of philosophy of science. This is why it strikes me as particularly important to take stock and systematize the philosophical approaches to models and to highlight the issues that arise in the context of models and theories. Due to the neglect of models in philosophy of science until comparatively recently, there exist no systematic accounts of the topic. An indication of this situation is the scarcity of monographs dedicated to the topic, the most notable exception being Mary Hesse's (1966) short book *Models and Analogies in Science.* There exists a small number of recent collections on the topic (Magnani, Nersessian, and Thagard 1999; Morgan and Morrison 1999; and Magnani and Nersessian 2002). Other than that, one finds a considerable number of articles on models spread across journals, sometimes in fairly obscure places. This is why I have included an extensive bibliography of the literature on scientific models at the end of the book. Not all works are referred to in the main text, although the more seminal are. This bibliography on scientific models covers the twentieth and the early twenty-first centuries. I hope that this bibliography is an aid to anybody who would like to pick up the thread laid out in this book and study further the philosophy—and the history of philosophy—of scientific modeling.

Notes

1. My understanding of a scientific model has an early precursor in that of Rosenblueth and Wiener 1945 (p. 316): "No substantial part of the universe is so simple that it can be grasped and controlled without abstraction. Abstraction consists in replacing the part of the universe under consideration by a model of similar but simpler structure. Models, formal and intellectual on the one hand, or material on the other, are thus a central necessity of scientific procedure." (I use "abstraction" in a different sense in chapter 6, section 6.5.)

2. For a more detailed discussion of this and related issues, see Böhm-Vitense 1992 or Tayler 1994.

3. In this sense the "standard model" of elementary particles is a theory.

4. Where the context allows, I shall from now on drop the "scientific" in front of "model."

References

Bailer-Jones, D. M. 2002. Scientists' thoughts on scientific models. *Perspectives on Science* 10: 275–301.

Bogen, J., and J. Woodward. 1988. Saving the phenomena. *The Philosophical Review* 97: 303–352.

Böhm-Vitense, E. 1992. *Introduction to Stellar Astrophysics, Vol. 3: Stellar Structure and Evolution.* Cambridge: Cambridge University Press.

Giere, R. 1999. *Science without Laws.* Chicago: University of Chicago Press.

Hesse, M. 1966. *Models and Analogies in Science.* Notre Dame, Indiana: University of Notre Dame Press.

Magnani, L., and N. J. Nersessian, eds. 2002. *Model-Based Reasoning: Science, Technology, Values.* New York: Kluwer Academic Publishers.

———, N. Nersessian, and P. Thagard, eds. 1999. *Model-Based Reasoning in Scientific Discovery.* New York: Plenum Publishers.

Morgan, M., and M. Morrison, eds. 1999. *Models as Mediators.* Cambridge: Cambridge University Press.

Rosenblueth, A., and N. Wiener. 1945. The role of models in science. *Philosophy of Science* 12: 316–321.

Tayler, R. J. 1994. *The Stars: Their Structure and Evolution.* 2nd edition. Cambridge: Cambridge University Press.

MECHANICAL MODELS 2

ACCORDING TO SOME SCHOLARS (for example, Jammer 1965, p. 167), the concept of a model started to be used in science in the second half of the nineteenth century. Models were used before that time, yet I take my point of departure from nineteenth-century models because these were the models to which twentieth-century philosophers of science mostly referred when they considered models in science.[1] Whether physically built or hypothetically conceived, early models were mechanical.

Being mechanical can mean two different things (cf. Schiemann 1997, pp. 21ff.):

1. Mechanical may mean that a process is explained in terms of the principles of classical mechanics (Newton, Lagrange); or
2. Mechanical may mean that motion is explained in terms of the "pushing" and "pulling" of things, whereby the motion is passed on through direct contact between the things involved (no action at a distance).

While the former has its roots in the *Philosophiae Naturalis Principia Mathematica* (1687) of Isaac Newton (1642–1727), the latter goes back to atomistic philosophies of Leucippus (fifth century B.C.) and Democritus (around 460–370 B.C.) and to the seventeenth-century mechanical philosophy of, for example, Pierre Gassendi (1592–1655),

René Descartes (1596–1650), and Robert Boyle (1627–1691). In what follows, I shall use "mechanical" roughly in the second sense, although with some further elaboration. If reference to the first sense needs to be made, I do so by talking about (belonging to) mechanics, meaning classical mechanics.

The above distinction between the two uses of "mechanical" is also related to what Thomas Kuhn (1977) has called the mathematical and the experimental tradition of science. The mathematical tradition focuses on principles and deduction, as exemplified in Newton's *Principia*, and corresponds to the previous first sense of "mechanical." The experimental tradition often relies on corpuscularian conceptions and therefore corresponds to the previous second sense of "mechanical." According to Kuhn's account, in antiquity the physical sciences of astronomy, optics, and statics all belonged more or less to the single field of mathematics. Data collection was uncommon, and these disciplines relied mostly on casual observations, besides mathematics. Kuhn names Galileo, Kepler, Descartes, and Newton as important seventeenth-century figures who all partook in this mathematical tradition of doing science and who at the same time had "little of consequence to do with experimentation and refined observation" (Kuhn 1977, p. 40).[2]

Kuhn contrasts this with experimental or "Baconian" science. The new experimental fields of research often had their roots in crafts and did not have the theoretical backup of the classical, mathematical sciences. Chemistry, for instance, had been the domain of pharmacists and alchemists. Mathematics was not perceived to play as much a role in these new areas of experimental investigation as it did in the classical sciences. The experimentalists were often amateurs with little mathematical skills, but adherents to corpuscularian ideas. Isaac Newton, as Kuhn highlights, presents an exception with regard to this separation in that he participated in and inspired both traditions: the mathematical tradition with the work of the *Principia* and the experimental, corpuscularian tradition with the work of the *Opticks, or a Treatise of the Reflections, Refractions, Inflections, and Colours of Light* (first published in 1704).[3] As a work about materials and their structure (and not just their motion and the production of motion by forces), the *Opticks* stood in the mechanical tradition of corpuscularian theories (Cohen 1956, p. 121). Thus it was not so much about mathematical principles.

Accordingly, the nineteenth century saw diverging trends, the mathematization of the experimental sciences on the one hand (section 2.1) and the employment of mechanical models on the other (section 2.2). Perhaps because models stood in stark contrast to the well-accepted mathematical trend, models prompted the need for explication or even promotion by their avid users. In section 2.3, I discuss some nineteenth-century considerations concerning the status of models, as they emerge from the writings of physicists who engaged in model building: models as promoting understanding and models as being not true or not real. Both mathematization and modeling took further turns in the twentieth century. Mechanical models nonetheless form the prototype of a scientific model in most people's imagination. It is therefore useful and necessary to consider the appeal of mechanisms even from today's perspective. Because physics has progressed, the concept of a mechanism has also evolved further, and in response to this, new notions of models being "mechanical" can be developed. I explore various recent conceptions of mechanism in section 2.4. Seeking the "essence" of mechanistic thinking in this way may forge a link between nineteenth-century and current scientific models. Current models may continue to be mechanical, albeit in a somewhat metaphorical or extended sense.

2.1 The Mathematization of Physics

When, at the beginning of the nineteenth century, Baconian science underwent mathematization (Kuhn 1977, pp. 71ff.), it was firmly tied into the tradition of Newtonian mechanics. Isaac Newton's mechanics had long been the ideal of successful physics because of the way in which it was laid out mathematically and in which it allowed the mathematical solution of an enormous range of physical problems. Besides Newton, the mathematization of physics is also an achievement of Leonard Euler (1701–1783), Jean d'Alembert (1717–1783), and others who undertook the task of developing Newtonian methods ("algebraizing" his work, if you will) (Simonyi 1995, pp. 294ff.) to produce detailed mathematical descriptions of different behaviors of rigid and elastic bodies and of ideal and viscous fluids.

While the eighteenth century had seen a mostly qualitative exploration of electrical effects, exemplified by the work of Joseph Priestley (1733–1804) or Benjamin Franklin (1706–1790) (ibid., pp. 323ff.), the onset of quantitative work was marked by the discovery of the force between charged bodies, Coulomb's law, that was discovered independently by a number of people—Joseph Priestley (1733–1804), Henry Cavendish (1731–1810), John Robinson (1739–1805), and Charles Auguste de Coulomb (1736–1806) (ibid., p. 329). Then, even the phenomena of electricity, magnetism, and heat were interpreted as closely as possible in the spirit of Newtonian mechanics. Coulomb's law, for instance, is put in a form similar to that of the law for gravitational attraction. In compliance with Newtonian mathematical ideals, the mathematical interpretation of phenomena became increasingly disjointed from the observable physical and geometrical properties of phenomena—that is, it became more and more abstract.

Jean Baptiste Joseph Fourier (1768–1830) demonstrated, with his mathematical treatment of heat, how to undertake a mathematical analysis of a phenomenon without querying its physical nature much, but purely based on experimentally tested laws (Harman 1982, pp. 27ff.). The success of this approach became further evident when William Thomson (1824–1907) successfully transferred Fourier's mathematical tools, originally devised for the phenomenon of heat, to quite a different area of application, when he in the 1840s elaborated a theory of electrostatics based on Fourier's theory of heat (Thomson [1842] 1942). The mathematical analogy between the treatment of heat and of electrostatics subsequently highlighted a physical analogy—that between thermal conduction and electrostatic attraction—where the distribution of electricity was represented by the flux of electrical force and the distribution of heat by the flux of heat (Harman 1982, p. 29). In the instance of heat and electrostatics, the mathematical treatment of the phenomena helped to overcome conceptual constraints by lifting the problem solution to the abstract and symbolic level of mathematical treatment.

Another route toward the mathematical treatment of a phenomenon is exemplified by James Clerk Maxwell's (1831–1879) vortex model. Maxwell exploited this mechanical model, taken from Michael Faraday's (1791–1867) ideas, to devise a mathematical description of the electromagnetic field. In his work *On Physical Lines of Force* (Maxwell [1861] 1890), he represented the electromagnetic field in terms of a fluid medium containing "vortices" (elastic cells) under a state of stress (Nersessian 1984, p. 77). The analogy runs roughly as follows: Maxwell imagined the magnetic field as a fluid filled with vortices that rotate with angular velocities that represent the intensity of the field and that are geometrically arranged in a way to represent the lines of force (Harman 1982, p. 89). As neighboring vortices that rotate in one direction would hinder each other, Maxwell introduced "idle wheels" between them and took these to correspond to the electric current. Maxwell then developed all his mathematical considerations from this model (Nersessian 1984, pp. 78–79) and was able to show that the field equations he derived with the help of it matched the experimentally observed electromagnetic effects. His strategy was, in his own words, "investigating the mechanical results of certain states of tension and motion in a medium, and comparing these with the observed phenomena of magnetism and electricity" (Maxwell [1861] 1890, p. 452).

In this example, mathematization and mechanization of physics are connected in the sense that a mechanical model of a phenomenon was mathematically described. This was so even though it could not be expected that the phenomenon to be considered (the magnetic field) would lend itself to description in terms of the mathematics of Newtonian mechanics. Yet, negotiated by the mechanical model, Newtonian mechanics could be exploited in such a way as to achieve a mathematical description of the magnetic field. That the mechanical approach continued to dominate theorizing even when the phenomena to be modeled seldom belonged to the field of classical mechanics is particularly intriguing.[4] Nancy Nersessian has commented: "What Maxwell did not realize is that the mechanical representation he wanted for the 'underlying processes' was not possible because in fact he had formulated the laws of a *non*-Newtonian dynamical system" (Nersessian 1984, p. 93).

This trend of sticking to concepts and mathematics of classical mechanics, instantiated in the form of mechanical analogies, clearly seemed to go against what the new physical problems suggested or required (Jammer 1965, p. 169). Electric and magnetic fields in particular could not easily be interpreted in terms of processes that consisted of the movement of corpuscles in time and space (which is why Newton's *Opticks* was so unpopular in those days; see Cohen 1956, p. 114), or they required the construction of a mechanical aether (see section 2.3). On the one hand, mechanical models could be used to obtain abstract mathematical interpretations. On the other hand, increasingly abstract and less mechanical phenomena seemed to be less and less suitable to being modeled mechanically, the most important example being electromagnetism. The continued construction of mechanical models by such physicists as Thomson and Maxwell must almost appear like a reaction to the fact that possibilities of modeling empirical phenomena in the familiar ways was on the wane. This kind of resistance shows, for example, in the comments Thomson made in his Baltimore Lectures: "I firmly believe in an electro-magnetic theory of light, and that when we understand electricity and magnetism and light, we shall see them all together as part of a whole. But I want to understand light as well as I can without introducing things that we understand even less" (Thomson [1884] 1987, p. 206).

To Thomson, Maxwell's mathematical theory appeared inadequate for understanding wherever it did not rest on mechanical models. Correspondingly, Thomson speaks of the "mathematical disease of aphasia," abstract formulae without concrete interpretations. This, he thought, required a cure and he praised Faraday for providing such interpretations—for example, by talking about "lines of force" (ibid., p. 148). Obviously, Thomson is right in requesting interpretations of abstract mathematical theories, but it is not obvious that they need necessarily be supplied by conventional mechanical models, nor was there any way of stopping the mathematization of physics itself (cf. Smith and Wise 1989, chapter 13, especially p. 445).

2.2 Scientific Models in the Mechanical Age

Throughout the nineteenth century, physicists brought the skill of developing mechanical models to unprecedented heights, and they did so not just as a route to mathematization. I cannot do justice here to the full range of mechanical models that nineteenth-century scientists developed, but I focus on some of the observations which such scientists as Maxwell, William Thomson (Lord Kelvin) (1824–1907), and Ludwig Boltzmann (1844–1906) made about models. The attraction of models, where recognized as such, is the attraction of explaining phenomena mechanically. I argue in section 2.4 that this attraction carries on to this day, albeit in altered form.

From today's perspective, it is not always easy to distinguish what we count as models from what was called a model in the nineteenth century. Both Maxwell's vortex model (section 2.1) and his model of electrical field lines as an incompressible fluid moving in tubes that he introduces in *On Faraday's Lines of Force* (Maxwell [1855] 1890) are what we would today call models. Maxwell, however, does not even mention the term "model." In this context he only addresses the issue of the use of analogy in science (see section 3.1). Boltzmann wrote an article on models for the tenth edition of the *Encyclopaedia Britannica*, reprinted in the eleventh edition, which is illuminating because it gives us a sense what a practicing scientist took a model to be. This article indicates that people took "model" to mean *physical* model. Boltzmann describes a model as "a tangible representation, whether the size be equal, or greater, or smaller, of an object which is either in actual existence, or has to be constructed in fact or in thought" (Boltzmann [1902] 1911, p. 638).

So although the thing that is modeled may or may not exist (yet), the model of it is something that physically exists. Correspondingly, the examples Boltzmann cites are models artists make before they produce a sculpture in an expensive or hard-to-manipulate material, such as stone or metal, or "plastic models" used in anatomy and physiology. In fact, physically building mechanical models was not an uncommon practice in the nineteenth century. It is further indicative that Boltzmann, who portrays models as a kind of language with which to express "our mechanical and physical ideas," claims that models "always involve a concrete spatial analogy in three dimensions" (ibid., p. 638). He goes on to emphasize potential disadvantages of a modeling approach in that models "are difficult to make, and cannot be altered and adapted to extremely varied conditions so readily as can the easily adjusted symbols of thought, conception and calculation" (ibid., p. 638).

Talking about "making models" is suggestive of physically building models rather than theoretically constructing them.[5] Moreover, it would be less of a problem to change and modify a theoretical construct than a model made from wood, metal, threads, cardboard, and so on. Models are precisely not theoretical entities such as "symbols of thought, conception and calculation." For Boltzmann the benefits of a model are related to enabling us to *see* things. This must mean that models were something that could be seen with one's eyes, not merely in the mind's eye. In view of the growing complexity of science, seeing as an efficient means of processing information had become indispensable, according to Boltzmann: "Yet as the facts of science increased in number, the greatest economy of effort had to be observed in comprehending them and in conveying them to others; and the firm establishment of ocular demonstration was inevitable in view of its enormous superiority over purely abstract symbolism for the rapid and complete exhibition of complicated relations" (ibid., p. 638).

Further, Boltzmann talks about "two cups or sheets fitting closely and exactly one inside the other" when modeling the refraction of light in crystals or about "a wave-surface formed in plaster [which] lies before our eyes" (ibid., p. 639). He also distinguishes between stationary and moving models. Moving models used in geometry "include the thread models, in which the threads are drawn tightly between movable bars, cords, wheels, rollers, &c." (ibid., p. 639). There is little doubt that Boltzmann took models to be literally tangible—that is, physical objects. His report concludes with the comment that the largest collection of models can be found in the museum of the Washington Patent Office (ibid., p. 640), so it is clear that he cannot take models to be theoretical constructs because these would be rather unsuitable for exhibition in that museum.

So the theoretical constructs, such as Maxwell's vortex model or his model of the electromagnetic field lines, which we call models today, were not models in nineteenth-century terminology. They are referred to as analogies. Boltzmann talks about "mechanical analogies" and then points out that mechanical analogies lead to the construction of mechanical models whereby "physical theory is merely a mental construction of mechanical models, the working of which we make plain to ourselves by the analogy of mechanisms we hold in our hands" (ibid.,

p. 639).[6] This distinction between model and analogy fits in not only with Maxwell's exclusive use of the term "analogy," but also with Pierre Duhem's (1861–1916) use of the terms "model" and "analogy" (Duhem [1914] 1954, pp. 95ff.).

With the rise in the use of models in science, a debate started about what the merits of models were, if any. Models were, for instance, seen as a road to visualization and understanding of abstract phenomena, especially when confronted with not obviously mechanical phenomena. In his Baltimore Lectures in 1884, William Thomson developed the wave theory of light by applying formulae of molecular dynamics and thus showed himself to be an ardent proponent of a mechanical world view. Correspondingly, he expressed his admiration for Maxwell's mechanical model of electromagnetic induction but was critical of Maxwell's more mathematical work. So why was it that Thomson attributed so much importance to mechanical models? A major reason is that he closely associated having a mechanical model with understanding a phenomenon: "It seems to me that the test of 'Do we or not understand a particular subject in physics?' is, 'Can we make a mechanical model of it?'" (Thomson [1884] 1987, p. 111).

This desire to understand by means of a mechanical approach is further stressed in another famous quotation from Kelvin's Baltimore Lectures—one that also contains the implicit criticism of Maxwell: "I never satisfy myself until I can make a mechanical model of a thing. If I can make a mechanical model I can understand it. As long as I cannot make a mechanical model all the way through I cannot understand; and that is why I cannot get the electromagnetic theory" (ibid., p. 206). Thomson does not tell us why he thinks that mechanical models promote understanding. He simply takes this for granted.[7]

While models proved extremely useful for the progress of physics, there remained the issue of what the status of models (or of analogues) was. For this, I review Maxwell's considerations with regard to the analogy between field lines and tubes in which an incompressible fluid moves in his *On Faraday's Lines of Force* (Maxwell [1855] 1890). Of course, as emphasized, Maxwell's example is an analogy and not a model in nineteenth-century terminology, but as Boltzmann highlighted, a mechanical analogy, being a "mental construction," can serve as an instruction for physically constructing a mechanical model.

Maxwell made it plain that electrical field lines were not tubes containing an incompressible fluid; this was "only" an analogy employed for the purposes of calculation. Maxwell expressed his view on the nonreality of the analogues in no uncertain terms: "The substance here treated must not be assumed to possess any of the properties of ordinary fluids except those of freedom of motion and resistance to compression. It is not even a hypothetical fluid which is introduced to explain actual phenomena. It is merely a collection of imaginary properties which may be employed for establishing certain theorems in pure mathematics in a way more intelligible to many minds and more applicable to physical problems than that in which algebraic symbols alone are used" (Maxwell [1855] 1890, p. 160).

Presumably, the same caveat would hold if a mechanical model was built on the basis of this analogy; then the model would still display some unrealistic properties, such as compression, that are not found in the phenomenon modeled. Two elements are relevant in employing the analogy. One is making the mathematics applicable to physical problems, the other is making it intelligible to the mind. The mind must be able to connect a physical phenomenon with a mathematical description: "In order therefore to appreciate the requirements of the science, the student must make himself familiar with a considerable body of most intricate mathematics, the mere retention of which in the memory materially interferes with further progress. The first process therefore in the effectual study of the science, must be one of simplification and reduction of the results of previous investigation to a form in which the mind can grasp them" (ibid., p. 155).

In short, such an analogue model is not seen as a description of how things really are, but as a mental tool, as providing a form in which the mind can grasp the intricate mathematics involved in dealing with a phenomenon. Just as Maxwell concerns himself with the reality of his models, Thomson considers their truth. Interestingly, even a model builder as avid as Thomson does not vouch for the truth of models. He says, while referring to a particular model of the molecular constitution of solids: "Although the molecular constitution of solids supposed in these remarks and mechanically illustrated in our model is not to be accepted as true in nature, still the construction of this kind is undoubtedly very instructive" (Thomson [1884] 1987, pp. 110–111).

If models are neither real nor true, the question arises to what extent their use in science is justified. Is it really enough if models serve certain heuristic and illustrative purposes, and nothing more? While Thomson viewed models, despite not being true, as indispensable, other scholars deemed such indispensability questionable (for example, Duhem; see chapter 4.3). Boltzmann summarized his own outlook on the justification of the use of models as a compromise: "At the present time it is desirable, on the one hand, that the power of deducing results from purely abstract premises, without recourse to the aid of tangible models, should be more and more perfected, and on the other that purely abstract conceptions should be helped by objective and comprehensive models in cases where the mass of matter cannot be adequately dealt with directly" (Boltzmann [1902] 1911, p. 638).

This almost sounds like models are to be tolerated until abstract characterization ultimately succeeds. The main reason for needing models, according to Boltzmann, is the "mass of matter"—that is, the sheer amount of material and information that needs to be dealt with. To capture this material, models are simple, more efficient means. The deciding circumstance is that the concrete and visible can be taken in more swiftly than the abstract and mathematical, but such appreciation of models as being concrete and visible is merely born out of human limitation in the face of complexity. Maxwell, in turn, who acknowledges the need for a concrete cognitively accessible conception, also highlights the need not to be carried away by a hypothesis that rests on an analogy: "We must therefore discover some method of investigation which allows the mind at every step to lay hold of a clear physical conception, without being committed to any theory founded on the physical science from which that conception is borrowed, so that it is neither drawn aside from the subject in pursuit of analytical subtleties, nor carried beyond the truth by a favourite hypothesis" (Maxwell [1855] 1890, p. 156). Thus a "clear physical conception" must be provided which is not purely abstract, but which is also not, as an analogue, unwarrantedly taken as real.

The scientists discussed in this section highlighted some of the benefits of models: promoting understanding, making mathematics intelligible, being instructive, summarizing an abundance of information by being visible. Yet as the nineteenth century drew to a close, there also existed a distinct unease connected with scientific models. This is evident in Boltzmann's position of sketching models as some kind of temporary or less-than-ideal solution to the practical problems of doing science. This trend gathered momentum in the early days of the twentieth century. Models, as understood in the nineteenth century as physical objects, were then at best thought to have temporary and heuristic advantages, being useful for cognitive and practical purposes, while a contrary widespread view was that they could even be eliminated in principle (see chapter 4). Many thought that the place of models could or should be taken by rigorous mathematical treatment and that this, although it may have emerged from models, would supersede them. In contrast to this, Thomson and Maxwell, although they do not claim reality nor truth for models, are firm about the mind's need for scientific models. This is an issue that did not receive attention again until much later in the history of the treatment of scientific models. Early glimpses appeared in the 1950s (see chapter 4, section 4.7), reinforced in the 1960s with Mary Hesse's work (see chapter 5, section 5.2), and more thoroughly appreciated only in the context of cognitive science (Bailer-Jones 1999).

2.3 The Mechanical Approach

Newton has made a lasting impact through his mechanics and the mathematics associated with it. But besides guiding the mathematical description of natural phenomena, as in the case of Maxwell's vortex model (section 2.1), thinking of processes as mechanical had other advantages for scientists. They were used to, and confident in, thinking in terms of mechanical processes that involved bodies acting on each other ("pushes and pulls"). However, there existed different approaches to what counted as mechanical. G. Schiemann (1997, pp. 92ff.) correspondingly distinguishes different traditions of mechanism: *materialist*, *dynamic*, and *dual* mechanisms. The materialist approach is typical of early corpuscularian positions—for example, that of Christiaan Huygens (1629–1695) or Robert Boyle (1627–1691). In it, an independent concept of force does not occur. Rather than forces at a distance, direct contact between corpuscles is the way motion gets passed on and natural phenomena are explained. The dynamic approach, associated with Gottfried Wilhelm Leibniz (1646–1716) and Immanuel Kant (1724–1804), in turn explains the properties of matter through forces alone and seems to have had no major impact on the development of physics at the time. In contrast, it was the Newtonian dual approach that came to dominate physics, especially in nineteenth-century modeling. It admitted both matter and force as fundamental. There remains the element of "push and pull" through direct contact between particles, but this is complemented by forces that can bring about motion even without direct contact through force fields (gravitational in the first instance, electrical and magnetic in the second).

According to Maxwell's article on *Physical Sciences* in the ninth edition of the *Encyclopaedia Britannica*, physics had indeed come to be understood as following a mechanical program based on "*the doctrine of the motion of bodies as affected by force*" (Maxwell 1885, p. 2; italics in original). "*The fundamental Science of Dynamics*" (ibid., p. 2; italics in original), consisting of different classes (kinematics, statics, kinetics, and energetics), was not conceivable without the concept of force, both kinetic forces acting between particles in direct contact and forces acting through force fields. Maxwell counterbalances this conceptual framework with emphasis on the experiential side of physical research, the Baconian tradition, and recommends that "in almost all of [the physical sciences] a sound knowledge of the subject is best acquired by adopting, at least at first, . . . the study of the connexion of the phenomena peculiar to the science without reference to any dynamical explanations or hypotheses" (ibid., p. 2).

In other words, the first step is the study of properties and empirical laws concerning phenomena. But then explanations should follow in the mold of the science of dynamics, given in terms of "the motion of bodies as affected by force." Typically enough, this method is thought not only to apply to what we now call "classical mechanics." Maxwell lists under the physical sciences for which dynamical theories can be developed the "(1) [t]heory of gravitation," the "(2) [t]heory of the action of pressure and heat" (thermodynamics), the "(3) [t]heory of radiance" (optics) and "[e]lectricity and magnetism" (ibid., pp. 2–3), while he admits that these sciences are in "very different stages of development" (ibid., p. 2). So the recommendation and the practice is to transfer a Newtonian dual mechanistic approach of bodies moving as affected by force to all other areas of physics.

Two examples serve to illustrate how this intrinsically mechanical approach penetrated nineteenth-century physics. One is the development of the kinetic gas theory, the other the pervasiveness of aether theories in the attempts to come to terms with electromagnetic phenomena. Maxwell himself contributed significantly to both these developments. The first important step for the conception of the kinetic gas theory was even to acknowledge that gases consist of particles, rather than the supposed heat-transporting fluid "caloric." Not until 1857 was the particle hypothesis revived, or rather, seriously considered. Rudolf Clausius (1822–1888), in his work *Über die Art der Bewegung, welche wir Wärme nennen,* which appeared in *Annalen der Physik* in 1857, proposed an atomic theory of heat arguing that heat was not an imponderable fluid or the like. Clausius linked pressure and heat to the motion of the atoms of which gases consist. Temperature became the average kinetic energy of atoms. Maxwell then refined the theory by considering the distribution of speeds and not just the simple average of speeds (Lindley 2001, p. 17). Boltzmann subsequently confirmed and elaborated Maxwell's approach. The result is now known as the Maxwell-Boltzmann distribution. When particles hit each other, they are acting as forces on each other—that is, they exchange momentum when they come into contact. Thus all elements of a mechanical system are present: particles, motion, (kinetic) force.

Interestingly, the effort of seeking an all-mechanical account of gases brought with it its own challenges in the case of kinetic gas theory—namely, the need for statistical laws. The number of particles and forces involved and the lack of specific coordinates for each particle made a statistical approach necessary. With this, the resulting laws could no longer be conceived as deterministic, as they would have been in the Newtonian framework. When Boltzmann argued in 1872 that entropy would always increase, his critics, among them Maxwell, pointed out that this was not necessarily so. This led Boltzmann eventually to concede that the increase of entropy was a matter of (very high) probability. Suddenly, the laws derived in a theory had become statistical, despite their mechanical roots.

The classical example of an ultimately unsuccessful adherence to mechanical explanations is the aether, for which a whole range of theories existed in the nineteenth century. Once the wave theory of light gained ground, light was imagined to be something like a vibration of a universal elastic medium called aether. Assuming that electromagnetic fields carry forces across distances through space, the reasoning was that something existed in space—that is, the aether—which must transport and pass on the force over a distance (Harman 1982, p. 72), rather like a sound wave is a pressure wave carried by matter. The aether is that which fills space and mechanically passes on the forces of the electromagnetic field, just as, for instance, in Maxwell's vortex model of the electromagnetic field (section 2.1). The aether was thought to consist of a continuum or of discrete particles, and in either case one part of the medium would act on its neighboring part to pass on the forces of the field (that is, by direct contact). Again the crucial elements are particles, motion, and force.[8]

The construction of more and more contrived "aethers" shows how difficult it had become to interpret field forces in terms of "push" and "pull." If the world was to be interpreted in mechanical terms, then what counts as mechanical could no longer be restricted to movement being passed on through direct contact ("push-pull"). Force had to be included as well. An aether based on contiguous particles failed increasingly to account for the properties of propagating electromagnetic waves. Similarly, gas particles make up the behavior of a gas by interacting fairly directly, but the behavior of the gas as a whole requires the concept of force. In a steam engine it is the pressure that is the motor behind the mechanical processes occurring in the machine, and pressure is the statistical behavior of particles, not the definite interaction of particles.

2.4 Mechanisms Since

With the advent of the theories of relativity and of quantum mechanics at the beginning of the twentieth century, mechanical interpretations of natural phenomena became yet more questionable—questionable with regard to both senses of "mechanical." The concepts of space and time were by the theory of relativity, and the notion of a push-pull mechanism by quantum theory, at least at the microscopic level. Neither relativity nor quantum mechanics lend themselves to straightforward mechanical interpretations, and they both employ very abstract and complex mathematical formalisms and techniques. Their formalisms and content do not always give rise to easily accessible mechanical models. In an article in *Philosophy of Science* in 1937, Philipp Frank described just this perceived superiority of the "mathematical conception of nature" over the "mechanical conception of nature": "It is seen today that nature can be described and understood not 'mechanistically' but only through abstract mathematical formulas. The world is no longer a machine but a mathematical formula" (Frank 1937, p. 41). An important issue in this is what one takes to be *anschaulich*, or intuitively perceptible (ibid., pp. 50f.).

Throughout the course of the twentieth century, consideration of scientific models has nonetheless been resumed, certainly by scientists and also by philosophers of science. This became possible partly because the concept of model changed. Models were no longer seen as having to be mechanical. Bohr's model of the atom, for instance, is both highly abstract and mathematical. Although some models have simply ceased to be mechanical, for others what it means to be mechanical has changed. For example, Rom Harré (1960) has specifically explored the link between models and mechanisms in an article called "Metaphor, Model, and Mechanism." The same text reappeared in *Theories and Things* (Harré 1961), entitled "Models to Mechanisms." Harré begins with what he calls principle P_1: "If you don't know why certain things happen then invent a mechanism (in accordance with the view you take of how the world works)—but it is better still if you find out how nature really works" (ibid., p. 18).

What scientists do is not necessarily discover mechanisms, but check hypotheses about mechanisms (ibid., p. 22). Harré calls these *hypothetical mechanisms* and equates them with models. It is a characteristic feature of science to accept or reject a hypothetical mechanism, or model, as the real mechanism (ibid., p. 25). Hypothetical mechanisms are necessary "because we have reached the limit of discernible mechanisms" (ibid., p. 27). This is precisely a reference to the way science has changed and to the way the concept of a mechanism may be changed with it.

Models that are mechanical provide mechanisms of what is going on as an observable empirical process. Mechanisms are known and familiar from machines, such as clockworks or bicycles. There is some indication, as I illustrate in this final section, that what a mechanism is can be understood in a more extended sense than was typical of the nineteenth-century context. So modern scientific models may be about mechanisms not literally in the sense of motion of bodies as affected by force, but in a somewhat different sense that still confers some of the benefits of mechanical models that Thomson thought of so highly. Indeed, there are at least two modern conceptions of mechanism (Glennan 1996 and 2002; and Machamer et. al. 2000) that show that they can do justice to some explanations and to discovery processes (Craver and Darden 2001) in modern science. I outline how they are, in certain senses, abstractions of the notion of motion of a body as affected by force. What remains in these modern conceptions of mechanism may be what needs to be considered to judge whether Thomson could have been right with his suggestion that models promote human understanding.

Stuart Glennan (1996) has explicitly made the point that his definition of a mechanism does not only apply to mechanical processes in the sense of belonging to classical mechanics. For example, a voltage switch can be analyzed as a mechanism in his sense. Correspondingly, his definition is on the one hand in the tradition of corpuscularian theories ("interaction of parts"), and on the other hand quite abstract: "A mechanism underlying a behavior is a complex system which produces that behavior by the interaction of a number of parts according to direct causal laws" (ibid., p. 52). When it is said that mechanisms are complex systems, "complex systems" is a generalization from what would be the "internal works of machines" (ibid., p. 51). Complex systems are interpreted analogously to machines. In a mechanism, it is *parts* that *interact.* Parts are important because Glennan emphasizes the capability of splitting up systems into parts (decomposition) and being able to analyze these parts separately—in other words to analyze bits of the behavior of the complex system independently of others. Which parts of a "machine"—that is, of a complex system—are to be considered depends on what it is exactly that the mechanism is supposed to explain. Comparing this with Maxwell's terminology, *parts* could be viewed as an abstraction of *bodies. Interaction* may be a combination of *motion* and *force*, or may correspond to *motion* alone, but belongs more to a corpuscularian framework than to Maxwell's dualist framework of mechanism.

Glennan proposes that a causal connection distinguishes itself from some accidental regularity in that a mechanism (producing the behavior of the system in question by the interaction of a number of parts) can be quoted that connects the two events that are thought to be causally connected. The existence of such a mechanism would lead to a *regular* connection between the two events and therefore to a causal law ("causal" because a mechanism can be given that connects the two events). Glennan highlights that the connection between the parts in the mechanism, the specific way in which these interact, need not be causal in order for the mechanism to establish causality. Thus causality between events is to be established at different levels; a causal explanation does not depend on a causal relation between the parts of its mechanism at a lower or even the fundamental level. It depends merely on saying how the parts of the mechanism interact, causal or not.

For fundamental laws, it is not possible to quote a mechanism; thus, according to Glennan, they are not causal. There are no causes for the workings of fundamental physics because, by definition, we cannot go deeper and highlight the mechanism that produces the corresponding fundamental laws. Glennan's example is the Einstein-Podolsky-Rosen (EPR) experiment. We cannot explain (and cannot provide a mechanism) why, if the spin of one photon in the experiment is known, then the spin of the other is also known. We can, however, use the knowledge of this fact and present this fact as a cause in explaining some higher-level process to which this fact may be instrumental. Consequently, mechanisms can be employed to explain higher-level laws, not fundamental laws. Glennan thus offers an account of causality that on the one hand fits an intuition linking causality with what is experienced with machines and on the other hand accommodates the fact that there are processes in nature that are not causal—that is, for which a mechanism cannot be given. The price to be paid is that causality does not feature as a fundamental relationship in nature; it instead depends on an understanding of mechanism.

The other competing conception of mechanism that I want to present is Peter Machamer, Lindley Darden, and Carl Craver's idea

(2000) in which mechanisms are interpreted in terms of *entities* and *activities*: "Mechanisms are entities and activities organized such that they are productive of regular changes from start or set-up to finish or termination conditions" (ibid., p. 3). Just like in Glennan's account, regularity is an important element in connection with mechanisms. Mechanisms should "work always or for the most part in the same way under the same conditions" (ibid., p. 3). Again, a crucial factor in a mechanism is that its elements can be individuated; Glennan would say that the mechanical process can be split into parts. There may be different stages to a mechanistic process, but they are all connected—something Machamer, Darden, and Craver refer to as *productive continuity*: "Complete descriptions of mechanisms exhibit productive continuity without gaps from the set up to termination conditions. Productive continuities are what make the connections between stages intelligible" (ibid., p. 3).

A mechanism is something that can be followed step by step, at least in our mind. For instance, one may understand how the wheels of a bicycle move by observing how the cyclist's feet exert a force first on one pedal and then on the other, which then rotate a rigid rod—that is, the axle, that is attached to a cogwheel in its middle. The pedals drive the cogwheel, over which a chain runs. This chain is connected to another cogwheel that is the axle to the back wheel of the bicycle. One movement gives rise to the next, and doing one thing causes another and thus sets a whole chain of events going. Of course, the vast majority of natural phenomena are not like the bicycle. They will not be so easily observable (for example, too big, too small, too fast, too slow) or the processes and changes that take place may not be as familiar and as easy to follow. However, some kind of extension from ordinary experience, even in the case of very abstract mechanisms, facilitates the mental exercise of following events in one's mind step by step and visualizing them as causally productive processes. Being given such an opportunity even in the case of "mechanisms" as they are encountered in modern science would, in my view, go some way in justifying Thomson's intuition that we only understand something if we can make a mechanical model of it.

Of course, it is not always possible to give complete descriptions of a mechanism. Productive continuity that is easily achieved when watching a bicycle—where all of the mechanism lies open to direct investigation—can be much harder to achieve during a discovery process. Because a mechanism is not necessarily discovered all at once, there may be gaps, which is why Craver and Darden (2001, p. 120) talk about "mechanism sketches." This highlights the intended continuity while there are still known gaps that need to be filled in. It indicates where more work needs to be done. By abstracting mechanisms, what were *bodies* for Maxwell become *entities*, and what were *forces* become *activities*, in this recent interpretation of mechanism. Activities of entities are *productive of change*, which may be like Maxwell's *motion*. The idea of regularity is already captured in Maxwell's reference to general laws that, in the elementary stage of science, are deduced from observed phenomena and are then analyzed in later stages (Maxwell 1885, p. 2).

The vocabulary that Machamer, Darden, and Craver (2000) chose to analyze mechanisms, in particular the notion of *activity*, suggest a link of mechanisms to ordinary experience that may make plausible the continuing use of mechanisms in science, which they have demonstrated for neurobiology and molecular biology. Activities are that which produce change, while entities are that from which and to which the changes happen. While *activity* is intended as a technical term, I cannot help thinking that it implies that there is an *agent* in the mechanism—as if there were little people running around doing things in the works.[9] This is not the disembodied machine that provided the blueprint for nineteenth-century scientific models, but one that is metaphorically enlivened. My image of little people running around is a caricature, of course. However, let me for a moment treat activities as a metaphor illuminating mechanisms. If there are empirical or theoretical entities that *act*, then we are encouraged to "empathize" with that entity and follow its *actions* mentally as if we were the actors inside the mechanism, in order to understand what is going on in a phenomenon.

Moreover, we are familiar with situations whereby many *agents*, each carrying out specific tasks—that is, engaging in different *activities* (Glennan's disembodied *interactions*)—can contribute to a common, overall activity. For instance, the writer writing, the editor editing, the printer printing, and so on are all active in producing a book; they work independently of each other and yet share a common project. The metaphor of activity thus illustrates that mechanisms that are activities are in some way combinatorial. Just as many agents may be involved in producing a book, so many mechanisms can contribute to and be part of accounting for an empirical phenomenon because they are "acting" synchronously, separately and yet together. The important elements of a mechanism are that things (bodies, parts, entities) can be individuated and separated within a phenomenon or process as they bring about change; that they are affected by force, interact, act—not only in a regular manner, but also in a way that can be followed and considered step by step, without any gaps in the chain. Perhaps models that are mechanical—in the sense that they satisfy these criteria of a mechanism—can really be said to make us understand.

2.5 Summary

Nineteenth-century models in physics were mechanical models. There are at least two meanings of "mechanical," one referring to the principles of classical mechanics in the tradition of Newton's *Principia* and one derived from corpuscularian theories and machines in the tradition of Newton's *Opticks*. The basic idea of the latter is of motion being passed on by direct contact or, extended to a dualist approach, to motion of bodies as affected by direct contact and by force.

Mechanical models assisted the mathematization of physics, as with Maxwell's vortex model, or served purposes of illustration, as with Faraday's lines of force, or both. The mechanical approach was applied widely, even to subject matters that do not belong to the subject area of classical mechanics—for example, to electromagnetism. There, the limitations of a mechanical approach became evident in the theories of the aether. The realization of these limitations continued into the twentieth century. Nonetheless, mechanical models were claimed by their proponents to provide understanding even if they were not real or not true. This association of a mechanical account with something that can be understood well has lived beyond the mechanical age. Although models today can certainly no longer be interpreted, for instance, in the spirit of corpuscularian theories or even in terms of Maxwell's view of physics as being about the motion of bodies as affected by force, there are contemporary conceptions of mechanism that preserve what is attractive about mechanical models and what it is about them that may make us understand. These conceptions involve the abstraction of the elements of the classical nineteenth-century mechanism. The strength of these conceptions of mechanisms is that they avoid the pitfalls of being conceptually restricted to the "motion of bodies as affected by force." In the light of such revised conceptions, it becomes possible to highlight the benefits of mechanical models for understanding without being retrograde about one's understanding of modern science.

Notes

1. For instance, Galileo Galilei (1564–1642) not only used models extensively in his scientific work, he also used the term *modello* in the second day of his *Dialogue Concerning the Two Chief Systems of the World* of 1632.

2. This is clearly a simplification given that Kepler, for example, systematically exploited Tycho Brahe's (1546–1601) observational data.

3. Certainly in the eighteenth century, Newton's work *Opticks* exerted remarkable influence on experimentalists and was read widely, something that was no longer the case in the nineteenth century, when the *Principia* was perceived as Newton's prime achievement (Cohen 1956, pp. 113ff.).

4. Boltzmann highlighted this already at the beginning of the twentieth century. He talked about the "mechanical theory of nature" and said about it the following: "In explaining magnetic and electrical phenomena [the mechanical theory of nature] inevitably fell into somewhat artificial and improbable hypotheses, and this induced J. Clerk Maxwell, adopting the ideas of Michael Faraday, to propound a theory of electric and magnetic phenomena which was not only new in substance, but also essentially different in form. . . . Maxwell propounded certain physical theories which were purely mechanical so far as they proceeded from a conception of purely mechanical processes. But he explicitly stated that he did not believe in the existence in nature of mechanical agents so constituted, and that he regarded them merely as means by which phenomena could be reproduced, bearing a certain similarity to those actually existing, and which also served to include larger groups of phenomena in a uniform manner and to determine the relations that held in their case" (Boltzmann [1902] 1911, p. 639).

5. Duhem ([1914] 1954, pp. 70ff.) gives a vivid and extended description of Thomson's model building as involving all sorts of materials.

6. Duhem also seems to indicate that a transition can be made from building mechanical models to imagining them and benefiting from them in that way: "In order to express in a more tangible manner the concrete character of the bodies with which he builds his mechanisms, W. Thomson is not afraid to designate them by the most everyday names: he calls them bell-cranks, cords, jellies. He could not indicate more clearly that what he is concerned with are not combinations intended to be conceived by reason, but mechanical contrivances intended to be seen by the imagination" (ibid., p. 75).

7. Duhem comments that Thomson, despite his predilection for models and "imaginative physics," has made his discoveries by means of "abstract systems" and that models, even in Thomson's physics, only represent results that are already obtained (ibid., pp. 97f.)

8. The Michelson-Morley experiment of 1887 is often taken to disprove the existence of an aether by showing that the speed of light is the same in the direction of Earth's orbit as perpendicular to it (Collins and Pinch 1993, pp. 29–42). If Earth were moving through a static aether, these speeds would be expected to be different. However, this alone is not necessarily enough to discount the aether. For instance, it might be postulated that Earth drags along the aether locally, in which case the speed of light would still be the same in all directions on Earth. More generally, it is really with the invariance of the speed of light in special relativity that the notion of the aether became increasingly obsolete. The crucial point is that if an aether were required to support light waves, then it must be possible to move relative to this aether, and therefore variations in the speed of light would be observed. In contrast, according to special relativity, the speed of light is invariant no matter what the frame of reference is—the Michelson-Morley experiment is precisely a way of attempting to observe an aether drift, that is, to measure a different velocity of light when moving relative to the aether.

9. Interestingly, John Herschel (1792–1871)—who talks about mechanisms in the context of finding out about the "hidden processes of nature" and the "discovery of the actual structure or mechanism of the universe and its parts"—also introduces the notion of "agents which are concerned in [the performance of the hidden processes]" (Herschel [1830] 1987, p. 191).

References

Bailer-Jones, D. M. 1999. Tracing the development of models in the philosophy of science. In L. Magnani, N. J. Nersessian, and P. Thagard, eds., *Model-Based Reasoning in Scientific Discovery.* New York: Kluwer Academic/Plenum Publishers, pp. 23–40.

Boltzmann, L. [1902] 1911. Model. *Encyclopaedia Britannica,* 11th ed., vol. 18. Cambridge: Cambridge University Press, pp. 638–640.

Collins, H., and T. Pinch. 1993. *The Golem.* Cambridge: Cambridge University Press.

Cohen, I. B. 1956. *Franklin and Newton.* Philadelphia: American Philosophical Society.

Craver, C. F., and D. Darden. 2001. Discovering mechanisms in neurobiology. In P. Machamer, R. Grush, and P. McLaughlin, eds., *Theory and Method in Neuroscience.* Pittsburgh: University of Pittsburgh Press, pp. 112–137.

Duhem, P. [1914] 1954. *The Aim and Structure of Physical Theory.* Translated from the French 2nd edition. Princeton, New Jersey: Princeton University Press.

Frank, P. 1937. The mechanical versus the mathematical conception of nature. *Philosophy of Science* 4: 41–74.

Glennan, S. 1996. Mechanisms and the nature of causation. *Erkenntnis* 44: 49–71.

———. 2002. Rethinking mechanistic explanation. *Philosophy of Science* (supplement) 69: S342–S353.

Harman, P. M. 1982. *Energy, Force, and Matter: The Conceptual Development of Nineteenth-Century Physics.* Cambridge: Cambridge University Press.

Harré, R. 1960. Metaphor, model, and mechanism. *Proceedings of the Aristotelian Society* 60: 101–122.

———. 1961. *Theories and Things.* London: Sheed and Ward.

Herschel, J. F. W. [1830] 1987. *A Preliminary Discourse on the Study of Natural Philosophy.* Chicago: University of Chicago Press.

Jammer, M. 1965. Die Entwicklung des Modellbegriffes in den physikalischen Wissenschaften (The development of the concept of model in the physical sciences). *Studium Generale* 18: 166–173.

Kuhn, T. S. 1977. *The Essential Tension.* Chicago: University of Chicago Press.

Lindley, D. 2001. *Boltzmann's Atom.* New York: The Free Press.

Machamer, P., L. Darden, and C. Craver. 2000. Thinking about mechanisms. *Philosophy of Science* 67: 1–25.

Maxwell, J. C. [1855] 1890. On Faraday's lines of force. In W. D. Niven, ed., *The Scientific Papers of James Clerk Maxwell, Vol. 1.* Cambridge: Cambridge University Press, pp. 155–229.

———. [1861] 1890. On physical lines of force. In W. D. Niven, ed., *The Scientific Papers of James Clerk Maxwell, Vol. 1.* Cambridge: Cambridge University Press, pp. 451–513.

———. 1885. Physical sciences. *Encyclopaedia Britannica*, 9th ed., vol. 19. Edinburgh: Adam and Charles Black, pp. 1–3.

Nersessian, N. J. 1984. *Faraday to Einstein: Constructing Meaning in Scientific Theories.* Dordrecht: Martinus Nijhoff Publishers.

Schiemann, G. 1997. *Wahrheitsgewissheitsverlust: Hermann von Helmholtz' Mechanismus im Anbruch der Moderne* (Truth-certainty loss: Hermann von Helmholtz's mechanism at the dawn of the modern age). Darmstadt: Wissenschaftliche Buchgesellschaft.

Simonyi, K. 1995. *Kulturgeschichte der Physik von den Anfängen bis 1990* (Cultural history of physics from the beginning until 1990). 2nd ed. Frankfurt: Verlag Harri Deutsch.

Smith, C., and M. N. Wise. 1989. *Energy and Empire: A Biographical Study of Lord Kelvin*. Cambridge: Cambridge University Press.

Thomson, W. [1842] 1872. On the uniform motion of heat in homogenous solid bodies. *Cambridge Mathematical Journal* 3. Also in *Reprint of Papers on Electrostatics and Magnetism*. London: Macmillan, pp. 1–14.

———. [1884] 1987. Baltimore lectures on wave theory and molecular dynamics. In R. Kargon and P. Achinstein, eds., *Kelvin's Baltimore Lectures and Modern Theoretical Physics*. Cambridge: MIT Press.

3 ANALOGY

THE OCCURRENCE OF ANALOGY has been closely associated with scientific modeling, as highlighted in the previous chapter. Scientific models are often based on analogies, and sometimes they are even said to *be* analogies. The topic of analogy arises almost naturally when models are discussed.[1] Moreover, some of the constructs called "analogy" in the nineteenth century would today be routinely referred to as "models," and it is reasonable to accept such an extension of the use of the term "model," as Mary Hesse (1953) has suggested. The cases to which this extension of the term "model" applies are those where models are not physically built, but where, for instance, there is an analogy in mathematical treatment of otherwise quite unrelated phenomena. Just think of the derivation of thermodynamic equations on the basis of classical mechanics that is supported by modeling a gas as billiard balls exchanging momentum.

Given that the topic of analogy often comes up in the context of modeling, we have to ask ourselves what the study of analogy might tell us about models. What is the insight we might achieve? The goal of this chapter is to learn about modeling by considering the role of analogy in science, assuming that drawing analogies has a share in building models. In section 3.1, I begin by considering what role scientists attributed to analogy. This takes me back again to the nineteenth century and the views of John Herschel (1792–1871), James Clerk Maxwell (1831–1879), Pierre Duhem (1861–1916), and Norman Robert Campbell (1880–1949). Then I turn to the study of the role of analogy in philosophy of science of the 1960s, when analogies were thought to have an explanatory function (section 3.2). All this gives us some indication of the connection between analogy use and scientific models. However, at this point it is also necessary to consider more formally what an analogy is. The structure-mapping approach to analogy, which is presented in section 3.3, distinguishes between attributes and relations in analogizing and provides the background for the computational study of analogy in cognitive science. Returning more specifically to the question of analogies and models, I give examples for the varied uses of analogies in science in section 3.4. I consider, in section 3.5, how analogy can be studied profitably. The lessons about how our insights concerning analogies contribute to understanding scientific modeling are presented in section 3.6.

3.1 Analogy Viewed from Science

Nobody would doubt that analogy can play a considerable role in arguments, for developing or for illustrating points. This use of analogy can be found in different intellectual pursuits, such as literature, historiography, or philosophy, and there exist many examples in science —some of which are presented in section 3.4. However, analogies were not just *used* in science; some scientists from the nineteenth century on also *discussed* their use.[2] John Herschel, mathematician, chemist, and astronomer, published a philosophical treatise in 1830 called *A Preliminary Discourse on the Study of Natural Philosophy.* In it he highlights the role of analogy in science. Herschel ([1830] 1987, p. 94) wrote that "parallels and analogies" can be traced between different branches of science, and doing so "terminate[s] in a perception of their dependence on some common phenomenon of a more general and elementary nature than that which form the subject of either separately."

In short, analogy enables us to link different branches of science, seeking something more general that joins them. Herschel's example of the search for and finding of such a common nature is electromagnetism, demonstrated by Hans Oërsted (1777–1851), relying on similarities between electricity and magnetism. Another example Herschel mentions is the analogy between light and sound. He concludes that, when confronted with an unexplained phenomenon, we need to find and study similar phenomena. If the causes are known that are efficacious in those other, similar phenomena, then this may give us clues concerning the causes in the phenomenon under consideration. Of course, whether this strategy works depends on the availability of closely analogous phenomena that are already explained (Herschel [1830] 1987, p. 148). Herschel (ibid., p. 149) wrote: "If the analogy of two phenomena be very close and striking, while, at the same time, the cause of one is very obvious, it becomes scarcely possible to refuse to admit the action of an analogous cause in the other, though not so obvious in itself."

The example Herschel gives for this kind of analogical inference is a stone being whirled around on a string on a circular orbit around the hand holding the string. Whoever is holding the string is able to feel the force pulling on the string due to the stone being kept on its circular orbit by the string. One may therefore, according to Herschel, infer that a similar force must be exerted on the moon by the Earth to keep the moon on its orbit around the Earth, even if there is no string present in the case of the moon and the Earth. Herschel obviously views this strategy of drawing analogies as a legitimate path toward scientific knowledge when he concludes from this example: "It is thus that we are continually acquiring a knowledge of the existence of causes acting under circumstances of such concealment as effectually to prevent their direct discovery" (ibid., p. 149).

The aim is to find the mechanisms that constitute the hidden processes producing certain phenomena (ibid., p. 191). In the course of this search, different hypotheses may be formed about the mechanism and, in fact, different theorists may well form different hypotheses about a phenomenon (ibid., p. 194), perhaps based on different analogies employed. Herschel is adamant that, despite the dilemmas that might occur due to competing hypotheses, forming hypotheses is still highly beneficial for scientific exploration: "they [hypotheses with respect to theories] afford us motives for searching into analogies" (ibid.,

p. 196). In particular, "it may happen (and it has happened in the case of the undulatory doctrine of light) that such a weight of analogy and probability may become accumulated on the side of an hypothesis, that we are compelled to admit one of two things; either that it is an actual statement of what really passes in nature, or that the reality, whatever it be, must run so close a parallel with it, as to admit of some mode of expression common to both, at least in so far as the phenomena actually known are concerned" (ibid., pp. 196–197).

Making such an inference is not only a theoretical undertaking, but it also has practical consequences. It leads to experiments: "[W]e may be thus led to the trial of many curious experiments, and to the imagining of many useful and important contrivances, which we should never otherwise have thought of, and which, at all events, *if* verified in practice, are real additions to our stock of knowledge and to the arts of life" (ibid., p. 197). In sum, analogies in science, according to Herschel, establish links between different areas of investigation. Moreover, they may aid explaining a new phenomenon on the basis of the causes acting in an analogous phenomenon already explained. As such, analogies lead to the formulation of hypotheses (and the stronger the analogy, the more evidence for the hypothesis). Finally, they may guide the experimental testing of such hypotheses, thus promoting theory development.

I discussed Maxwell's vortex model, his model of field lines, and mechanical understanding of physics in the last chapter. I now specifically consider his reflections on the use of analogy in science. One place where he explicitly considered the use of analogy in science is his "Address to the Mathematical and Physical Sections of the British Association" in Liverpool in 1870 (Maxwell [1870] 1890).[3] He had in mind what we would today call a cognitive function of analogy. Maxwell's emphasis is on the illustrations provided by analogies. The scientist should "try to understand the subject by means of well-chosen illustrations derived from subjects with which he is more familiar" (ibid., p. 219). The aim is "to enable the mind to grasp some conception or law in one branch of science, by placing before it a conception or a law in a different branch of science, and directing the mind to lay hold of that mathematical form which is common to the corresponding ideas in the two sciences, leaving out of account for the present the difference between the physical nature of the real phenomena" (ibid., p. 219). An analogy is more than a teaching aid or dispensable illustration and can tell us what the system under study is like: "the recognition of the formal analogy between the two systems of ideas leads to a knowledge of both, more profound than could be obtained by studying each system separately" (ibid., p. 219).

Maxwell voices reasons why analogies are useful in science. Using an analogy can be a strategy toward understanding an as yet not understood subject and hence contribute to generating knowledge by guiding the mind. Understanding, knowledge generation, and creativity are issues that have been investigated quite systematically in the cognitive sciences. Maxwell was also a great practitioner of the use of analogy in physics. Analogy was deliberately and systematically used in an effort to understand electrical effects that were not mechanical by nature. (Frequently these were analogies to mechanical models, see chapter 2, section 2.1, but not inevitably.) Maxwell not only devised the vortex model sketched in chapter 2. An earlier and very instructive example, mentioned in section 2.2 of that chapter, is Maxwell's model of electrical field lines that quite obviously exploits an analogy (Maxwell [1855] 1890). There, the field lines are represented by tubes in which an incompressible fluid moves, a model also broadly based on Thomson's analogy between the flow of heat and the flow of electrical force. Maxwell's aim in using this analogy is not to say what electrical field lines are physically like—they do *not* consist of a fluid—but how one might imagine them in accordance with the analogy guiding their theoretical description.

Maxwell (ibid., pp. 157–158) wrote: "By the method which I adopt, I hope to render it evident . . . that the limit of my design is to show how, by strict application of the ideas and methods of Faraday, the connexion of the very different orders of phenomena which he has discovered may be clearly placed before the mathematical mind." Here Maxwell makes explicit how analogies, and analogies to classical mechanics in particular, aid understanding. One beneficial point is that "the mathematical forms of the relations of the quantities are the same in both systems, though the physical nature of the quantities may be utterly different" (Maxwell [1870] 1890, p. 218).

Exploring an analogy at the level of mathematical description not only inspires a mathematical description, but can also aid the interpretation of this new mathematical description, as in the elaboration of William Thomson's analogy between electrostatics and heat flow (see chapter 2, section 2.1).[4] Maxwell very much saw analogy as a mental aid instructing us how to imagine things, or guiding us how to come to terms with a mathematical description. He also acknowledged that analogy can be a path to knowledge, even to some kind of "more profound" knowledge, recognizing parallels between different branches of empirical investigation.[5] This is somewhat similar to Duhem, who takes analogies to exist as formal relationships between phenomena or, rather, between the theoretical treatment of phenomena. Pointing to examples, such as light and sound waves or magnetism and dielectric polarization, Duhem ([1914] 1954, p. 96) stressed that "it may happen that the equations in which one of the theories is formulated is algebraically identical to the equations expressing the other. . . . [A]lgebra establishes an exact correspondence between [the theories]."

To find such a correspondence serves "intellectual economy," and it can also "[constitute] a method of discovery" by "bringing together two abstract systems; either one of them already known serves to help us guess the form of the other not yet known, or both being formulated, they clarify each other" (ibid., p. 97). There is an element of the cognitive side of analogy use which Maxwell described so vividly, but Duhem is not fond of these cognitive elements. (See the more detailed discussion of Duhem in chapter 4, section 4.3.) Duhem asserts that there is "nothing here that can astonish the most rigorous logician"; it is a strategy in perfect agreement with "the logically conducted understanding of abstract notions and general judgements" (ibid., p. 97). Still, he judges analogies in science as heuristic, thus as no longer essential once a theory is formulated.

Campbell ([1920] 1957), another physicist discussing analogy, has quite a different approach to this subject. In his account a theory contains a hypothesis, a set of propositions of which it is not known whether they are true or false of a certain subject. (I call this the "first subject.") If the ideas expressed in these propositions were not connected to some *other* ideas (associated with a "second subject"), they would be, according to Campbell, no better than arbitrary assumptions.

As the propositions constituting the hypothesis are not themselves testable, they require some kind of confirmation via a translation into *other* ideas—that is, ideas about a second subject—whereby the latter are known to be true through observational laws. This is the analogy. Campbell's example is the kinetic theory of gases (first subject) being the hypothesis formulated in terms of a set of propositions. The second subject would then be "the motion of a large number of infinitely small and highly elastic bodies contained in a cubical box" (ibid., p. 128). For the latter, laws are available so that it can be known which propositions concerning the second subject are true. The first subject, Campbell proposes, benefits from this knowledge. Via a "dictionary" the transition from first to second subject can be made, and the knowledge about the small elastic bodies illuminates the case of interest —that is, the first subject—and is employed to test the hypothesis indirectly. Campbell (ibid., p. 129) asserts: "In order that a theory may be valuable . . . it must display an analogy. The propositions of the hypothesis must be analogous to some known laws." In modern terms, the small elastic bodies in the box would be considered as a model, although Campbell does not use this term.

It is important to note that Campbell does not advance analogy in the way so far encountered with Herschel or Maxwell. For him, analogy is the relationship between a theory and something to which known observational laws apply. As D. H. Mellor (1968) has rightly observed, Campbell requires analogy largely to overcome the abyss between theory and observation.[6] Campbell is little concerned with the advantages of analogy for the practice of science—that is, for discovery or teaching. Quite to the contrary, he emphasizes that his reason to attribute crucial importance to analogy in science has nothing to do with analogies being tools in the process of discovering and developing theories. "[I]t seems to me," Campbell wrote, "that most of [the systematic writers on the principles of science] have seriously misunderstood the position. They speak of analogies as 'aids' to the formation of hypotheses (by which they usually mean what I have termed theories) and to the general progress of science. But in the view which is urged here analogies are not 'aids' to the establishment of theories; they are an utterly essential part of theories, without which theories would be completely valueless and unworthy of the name. It is then often suggested that the analogy leads to the formulation of the theory, but that once the theory is formulated the analogy has served its purpose and may be removed and forgotten. Such a suggestion is absolutely false and perniciously misleading" (Campbell [1920] 1957, p. 129).

The polemic goes on, talking of absurdity and perversity. So it would be wrong to portray Campbell as an early defender of the role of analogy in science in the cognitive, creative sense that most would now associate with the topic. According to him, the point of a theory is not only that it explains empirical laws by logical deduction, but that a theory is chosen because it displays an *analogy* to the empirical laws in question. Correspondingly, Campbell asserts with regard to this second, heuristic sense of analogy use: "Analogy, so far from being a help to the establishment of theories, is the greatest hindrance. It is never difficult to find a theory which will explain the laws logically; what is difficult is to find one which will explain them logically and at the same time display the requisite analogy" (ibid., p. 130). Analogy, presenting the factual relationship between theory and empirical laws, is even referred to by Campbell in terms of truth or falsity (ibid., p. 130).[7] So it has certainly nothing to do with hypothesis generation, illustration, or a "thinking aid."

Considering analogy, Herschel and Maxwell make related points, yet with quite different emphases. Knowledge generation, and gaining some deeper knowledge, via analogy is a common and important theme. Duhem thinks that analogy is not much more than a "method of discovery" to be discarded after use, whereas Campbell places more emphasis on needing analogy, albeit not for purposes of discovering theory. Herschel sees analogy as leading to the formulation of a hypothesis, besides linking different areas of investigation. In addition, he pays attention to the potential of both reasoning and discovery by analogy. Trying out analogies thus becomes a sensible research strategy. Finally, Maxwell goes beyond these obvious uses of analogy in practice. According to him, analogy can be used for illustration, relating something unfamiliar to something familiar; it instructs us how to imagine something and, in particular, how to come to terms with highly abstract mathematical expressions when applied in a new branch of physics, and this is more than just the "intellectual economy" that Duhem grants.

3.2 The Explanatory Function of Analogy

Having considered the views of what a number of scientists took to be the role of analogy in science, let me continue with philosophers' treatment of this question. For reasons to be explored in chapter 4, models and analogies were not a topic attracting attention in the philosophical climate reigning from the 1920s on. In the 1950s at the earliest, these topics reentered the philosophical debate.

The issue of theory development, the exploration of theoretical approaches to phenomena, provides the stage on which Hesse ([1963] 1966 and 1967; see also Harré 1970 and 1988) famously distinguishes between positive, negative, and neutral analogy. The positive analogy refers to those things that two analogues are known to have in common; known differences are referred to as negative analogy; all those properties for which commonalty or difference is yet to be established are referred to as neutral analogies. The idea behind this is that the neutral analogy often indicates that something is a suitable subject for further examination, and in that sense applying the analogy as a whole can serve as a guide to further investigation. In her dialogue between a Campbellian and a Duhemian, Hesse puts the following words in the mouth of the Campbellian talking about the billiard ball model of the "dynamical" theory of gases:

> Let us call those properties we know belong to billiard balls and not to molecules the *negative analogy* of the model. Motion and impact, on the other hand, are just the properties of billiard balls that we do want to ascribe to molecules in our model, and these we can call the *positive analogy*. Now the important thing about this kind of model-thinking in science is that there will generally be some properties of the model about which we do not yet know whether they are positive or negative analogies; these are the interesting properties, because, as I shall argue, they allow us to

> make new predictions. Let us call this third set of properties *neutral analogy*. If gases are really like collections of billiard balls, except in regard to the known negative analogy, then from our knowledge of the mechanics of billiard balls we may be able to make new predictions about the expected behavior of gases. Of course the predictions may be wrong, but then we shall be led to conclude that we have the wrong model. (Hesse 1966, pp. 8–9)

It is clear from this that Hesse's main concern is with the potential of a model to grow and develop, a model being "the imperfect copy (the billiard balls) *minus the known negative analogy*" (ibid., p. 9). In a second sense of the term "model," the model may include the negative analogy, although the point is that it always includes the neutral analogy. The main distinction from theory seems to be that theories only include the positive analogy, neglecting the neutral aspects of the analogy that might have the potential for prediction (ibid., p. 10). The emphasis of this use of analogy is on discovery, exploration, and prediction.

One reason why analogy is often thought to occur in science is because it supports a central function of models: explanation (Nagel [1960] 1979, pp. 107ff.; Harré 1960; Hesse 1966; and Achinstein 1968).[8] According to some authors, models being explanatory mostly coincides with them being developed on the basis of an analogy to some other known object or system (Achinstein 1968, p. 216). Explanation is thus linked to making the transition from something unfamiliar to something more familiar: "The analogies help to assimilate the new to the old, and prevent novel explanatory premises from being radically unfamiliar" (Nagel [1960] 1979, p. 46).

Analogy counts as a plausible candidate for providing explanations because the use of more familiar and already accepted models (models that have led to understanding in different but comparable situations) appears as a sensible strategy in a new context. Correspondingly, Peter Achinstein (1968, pp. 208–209) has stated: "Analogies are employed in science to promote understanding of concepts. They do so by indicating similarities between these concepts and others that may be familiar or more readily grasped. They may also suggest how principles can be formulated and a theory extended: if we have noted similarities between two phenomena (for example, between electrostatic and gravitational phenomena), and if principles governing the one are known, then, depending on the extent of the similarity, it may be reasonable to propose that principles similar in certain ways govern the other as well."

Achinstein quotes as examples, among others, the analogies between the atom and the solar system; between waves of light, sound, and water; between nuclear fission and the division of a liquid drop; between the atomic nucleus and extranuclear electron shells; and between electrostatic attraction and the conduction of heat (ibid., pp. 203–205). Achinstein does not, of course, discuss under which conditions it *is* reasonable to assume that principles holding in one phenomenon also hold in another. It is important to note that Achinstein, and probably others in the 1960s, not only think of analogy as explanatory, but also treat it as very closely associated with (if not identical to) scientific models. I think this is somewhat misleading, however. In the next section, I discuss what an analogy is and how it is different from a model, as "model" was introduced in chapter 1.

Very roughly, an analogy is a relationship, whereas a model is a (interpretative) description of a phenomenon, whereby this description may or may not exploit an analogy. Although some models may be seen as having their roots in an analogy, such as J. J. Thomson's plum pudding model of the atom or Bohr's model of the atom based on the planetary system, few existing models in science have not developed beyond the boundaries of the analogy from which they originated. If anything, a model could be an analogue, but this is not entirely the issue because the way to evaluate a model is not to judge whether it is analogous to something, but whether it, as it stands (analogous or not), *provides access* to a phenomenon in that it interprets the available empirical data about the phenomenon in a certain way. Rather than *be* a model, an analogy used for modeling can act as a catalyst to aid modeling and provide ideas in the course of a modeling effort. It is therefore more than reasonable to stress the importance of analogy in the modeling process, given that analogy is one of the cognitive strategies available for creative discovery from which scientific models result (see also chapter 5).[9]

3.3 Attributes and Relations in Analogies

Before more can be said about what analogy has to do with models, it is time to consider more precisely what an analogy is. For this I draw from the logical analysis of analogy by Mary Hesse, on the one hand, and from the not too dissimilar structure-mapping approach to analogy by Dedre Gentner, a cognitive psychologist. The Greek word "analogy" means "proportion"—for example, 2 is to 4 as 4 is to 8. Analogy is often understood as pointing to a resemblance between relations in two different domains—that is, A is (related) to B like C is (related) to D. Modern uses of the term "analogy" mostly go well beyond the narrow meaning of proportion. To use a current example, the electrons in an atom are related to the atomic nucleus like the planets are related to the Sun. But how is this analogical relationship to be analyzed exactly?

Hesse introduces a distinction that I want to outline here, the distinction between formal and material analogy (Hesse [1963] 1966, pp. 57ff.; see her "Essay on 'Material Analogy'"). The term "formal analogy" points to cases where the relations between certain elements within one domain are identical, or at least comparable, to the relations between corresponding elements in some other domain. Such an identity of structure does not require a "material analogy"—that is, the individuals of the domains ("things") are not required to share attributes, or properties (Hesse 1967). Material analogies, in turn, are based on two things or analogues having certain properties in common. Hesse assumes that these properties are observable.[10] Her example is the analogy between the Earth and the moon, which have properties in common, such as being large, solid, opaque, spherical bodies receiving heat and light from the Sun, revolving around their axes, and being attracted by other bodies (Hesse 1966, p. 58). As such, material analogies are pretheoretic analogies between observables.

Hesse draws a further distinction between whether the properties of the two analogues are identical or merely similar. The common properties for the moon and the Earth which Hesse lists are identical for both the Earth and the moon. If one, however, compares the properties of light and sound, then these are not identical, although they can be viewed as similar: sound echoes and light reflects; sound is loud, whereas light is bright; sound has pitch and light has color; sound is detected by the ear, whereas light is detected by the eye (ibid., p. 60). Hesse draws tables in which she lists the properties in vertical columns next to each other. This is why she refers to corresponding properties (similar or identical) of two analogues as "horizontal relations." She also considers the "vertical relations." These are relations between the properties of one analogue—for instance, causal relations between sound, echoing, being loud or quiet, having pitch and being detected by the ear. Of course, it is possible that similarities (or identity) exist between the vertical relations of two analogues, and this is the formal analogy.

As far as the vertical relations are concerned, there are a number of different ways in which they might be connected. Sometimes it may even be hard to determine the type of relation. For instance, Hesse introduces the example of an analogy between birds and fish, considering wing as analogous to fin, lungs to gills, and feathers to scales (ibid., p. 61). These are analogous with regard to their function. Wings and fins allow birds and fish, respectively, to move, lungs and gills to breath, and feathers and scales to be protected from the influences of the immediate environment. So, with regard to the horizontal relation, one could propose a similarity of functional relation. It gets harder when looking for a relationship between wings, lungs, and feathers. They do not seem to be causally related, other than perhaps in a distant evolutionary sense. Their vertical relation only seems to consist in being parts of the same whole—namely, the bird. Especially concerning the status of the similarity between bird body parts and fish body parts, this example illustrates how complicated the analysis of an analogy can be.

In Hesse's account of scientific examples, a material analogy seems to be that which provides the basis for finding out about a formal analogy. A nonscientific example that Hesse uses for illustrating formal analogy is the analogy suggesting that the state is to citizens as a father is to children (ibid., p. 62). In this case, there are no observable properties of note that state and citizens have in common with father and children. The only similarity conceivable between the two analogues is that of vertical relations, such as providing for, protecting, being obedient to. Hesse claims that stating these vertical relations is no basis for prediction, presumably because "state" and "father" and "children" and "citizens" have no properties in common.[11] Correspondingly, Hesse suggests that prediction is only feasible if there is some horizontal similarity between the corresponding terms of the analogues. This is a tricky point to argue, partly because similarity is such a slippery concept. Maybe father and state have some observable properties in common after all, and maybe some predictions based on analogy work out, even though the similarity between the two domains is minimal.

Hesse's general point is correct, however. An argument from vertical analogy is never as strong as an argument from property induction. She writes: "The difference between the two types of argument is due to the amount of information available, and analogical argument is necessary only in situations where it has not been possible to observe or to produce experimentally a large number of instances in which sets of characters are differently associated" (ibid., pp. 75–76). In other words, analogical argument may be available in cases where straightforward generalization is not. This is perfectly acceptable given that the goal of analogy is not the same as that of induction. An analogy needs to be suggestive. Formally speaking, the goal of using analogy is to help to select hypotheses (ibid., pp. 76–77).

Vertical relations may be causal "in some acceptable scientific sense" (ibid., p. 87). One type of similarity between two object analogues might be that the same causal relations exist between their properties. For instance, if sound and light are treated as analogues, then one might say that pitch and color "amount to the same thing" in the context of their respective theories. Also, phenomena such as the Doppler effect exist for both in a comparable way. It remains unclear to me whether the formal and the material analogy between two things can always be separated clearly in the process of developing a model,[12] or whether mixing material and formal elements in the use of an analogy cannot also be a stage in the (advanced) discovery process that produces further insight.

Another approach to analogy comes from cognitive psychology. Gentner (1982 and 1983) developed a whole framework for analogy: analogy as structure mapping.[13] This is, as far as I can see, another use of the idea of formal versus material analogy. The first important point Gentner makes is that an analogy does not become stronger only because the two (material) analogues have more properties in common. Rather, some properties are more relevant than others to making a strong analogy. She writes: "A theory based on the mere relative number of shared and nonshared predicates cannot provide an adequate account of analogy, nor, therefore, a sufficient basis for a general account of relatedness" (Gentner 1983, p. 156).

First of all, Gentner distinguishes between attributes and relations. Attributes are predicates that take one argument, such as LARGE(x), whereas relations are predicates that take two arguments, such as COLLIDE(x,y). She views drawing an analogy as a mapping between two domains, the *target* and the *base domain.* The base serves as a source of knowledge for the target that is the domain to be explicated. The aim is to check out potential inferences that are drawn about the target domain, and for this purpose, predicates from the base domain are "carried across" to the target domain. So although the point of departure (actual human reasoning) is different from Hesse's (logical validity of analogy inferences), the phenomenon examined (the potential and the conditions of drawing inferences based on analogy) is the same.

The emphasis is on carrying across relations rather than attributes, but analogy is not the only way two different domains can map onto each other. Gentner characterizes different types of domain mappings as follows: "Overlap in relations is necessary for any strong perception of similarity between two domains. Overlap in *both* object-attributes and inter-object relationships is seen as literal similarity, and overlap in *relationships* but not objects is seen as analogical relatedness. Overlap in *object-attributes* but not relationships is seen as a mere appearance match. Finally, a comparison with neither attribute overlap nor relational overlap is simply an anomaly" (ibid., p. 161). One claimed result of this analysis is that "the contrast between analogy and literal similarity is a continuum, not a dichotomy"(ibid., p. 161).

How well an analogy works also depends on the analogous predicates being interrelated with each other (Hesse's vertical relations), forming a whole system of connected knowledge. Gentner (ibid., p. 162) writes: "Part of our understanding about analogy is that it conveys a system of connected knowledge, not a mere assortment of independent facts." She correspondingly introduces her *systematicity principle*: "A predicate that belongs to a mappable system of mutually interconnecting relationships is more likely to be imported into the target than is an isolated predicate" (ibid., p. 163). Her example is the solar system and the Rutherford model of an atom. In this case, predicates —such as the distance from Sun to planet or from nucleus to electron, the attractive force between the two, and how massive or how charged the objects (Sun and planet, nucleus and electron) are—are related to each other in that changing one affects the others.[14] Empirical evidence for the systematicity claim can, for instance, be found in Clement and Gentner 1991 or in Markman 1997.[15]

So, in contrast to some historical or philosophical approaches, the structure-mapping analysis of analogy is founded on empirical results of psychological research. While Hesse approached analogy through a logical analysis, her results are fundamentally in line with Gentner's. Relations within a domain, and transferred to another domain, are central in rating analogies, and analogies are considered more apt when a lot of relational information is available, rather than merely shared attributes. Hesse's and Gentner's work provides us with a working analysis of what analogy is. I now turn to the ways in which analogies are used in science, to make some progress on the question of what analogy and models have to do with each other. After all, we want to learn what insights we can gain from analogy for our study of models.

3.4 What Analogies Do in Science

If analogies are not models—analogies being relations and models being descriptions—then analogies may still be a "mechanism" of modern scientific thinking.[16] There can be no doubt that analogies are widely used in science. In this section, I concentrate on distinguishing various uses to which analogies are put in science. Analogies are sometimes instrumental in scientific discovery; they are often used for the development of ideas and scientific accounts. They can guide experimental design, they may influence the evaluation of a scientific account, and are frequently employed in instruction and for illustration. No one analogy will be used for all these purposes, but perhaps for one or two. Correspondingly, the kinds of uses of analogies encountered in the sciences differ quite significantly. Moreover, although it is comparatively easy to claim that analogies play *some* role in scientific discovery, science teaching, and so on, it is much harder to determine what this role is exactly and how crucial exactly it is for doing science. This is why I also consider, in section 3.5, what our options are for the study of analogy use in science.

The body of examples that encouraged associating models with analogy comes from nineteenth-century physics (cf. Duhem [1914] 1954; Campbell [1920] 1957; and Hesse 1953). As spelled out earlier, examples include the analogy between heat and electrostatics (North 1980, pp. 123ff.) or Maxwell's approach to electromagnetism by analyzing an electromagnetic aether in terms of vortices along lines of magnetic force (see chapter 2, section 2.1). It is noticeable that the analogies that Maxwell, Duhem, and Campbell referred to were analogies in the mathematical treatment of empirical phenomena, and I refer to them as mathematical analogies.[17] Maxwell already identified this analogy use where "the mathematical forms of the relations of the quantities are the same," but the physical nature of the quantities different. The point of mathematical analogies is that the same, or a very similar, mathematical formalism is applied in two different domains. It can be the case that for a certain source domain—for example, the mathematical description of heat phenomena—a mathematical apparatus already exists and is well elaborated and is then transferred to a different domain, or two domains develop relatively independently and similar equations can then be formulated by writing down a law in a particular way. An example is the equations for gravitational force and for electrostatic force (Coulomb's law), where a law in the style of Newton's force law was specifically sought:

$$F_{grav} = G\frac{m_1 m_2}{r^2} \text{ and } F_{el} = \frac{1}{4\pi\varepsilon_0}\frac{q_1 q_2}{r^2}.$$

The most important parallel here is that both the gravitational and the electrostatic force are proportional to the inverse of the square of the distance, r, between two masses, m_1 and m_2, or two charges, q_1 and q_2. Everything else is physical constants that scale the quantities and ensure that the units come out right. It is striking that the equation for these two force laws displays a strong analogy, while the physical processes that are acting—gravitational and electrostatic attraction—are physically different. This makes this case a good example of Gentner's structure-mapping approach or Hesse's formal analogy (section 3.3). Charge and mass have few, if any, attributes in common, but a relation can be found in both domains, expressed by the force law, in which the charges and the masses play an analogous role. Moreover, being able to formulate such an analogous equation can be taken as a sign of linking two domains and providing understanding in the way that Herschel claimed.

It is one thing if the transference of analogous equations is "found" in nature when two descriptions arrived at independently are found to be analogous. It is another if a mathematical framework is transferred to a domain of technical application. Simulated annealing is an optimization technique that aims to find the optimal solution to a multidimensional problem (Bailer-Jones and Bailer-Jones 2002). Multidimensional means that many different variables are involved in the solution of the problem, which makes it impossible simply to draw a plot in two or three dimensions and to inspect it visually. Such techniques are needed in any area of science when having to cope with large complex data sets. The optimization technique of simulated annealing is constructed in analogy to the physical process of annealing when a material is heated to a very high temperature and then slowly cooled.

The material is cooled slowly to encourage it to solidify into its lowest energy state, corresponding to a single crystal. If a material is cooled very rapidly ("quenching"), the material is therefore likely to become "frozen in" at a metastable, fairly high energy state—for example, one corresponding to a polycrystalline material. Slow cooling is proposed to be like the search for an optimal solution (that is, a low energy state), which is why the whole mathematical apparatus from statistical mechanics that is applied in the case of real annealing can be transferred to the optimatization problem occurring with large data sets. The analogy employing the Boltzmann formula is so close that the whole conceptual background of energy, temperature, entropy, heat capacity, and so on is also transferred to the domain of optimization by means of simulated annealing. While this is a strong mathematical analogy, just like some analogies encountered in physics, it is not one that serves the description of nature, unlike the equations for gravitational and electrostatic force. The analogy helps to develop a data processing technique that, in turn, can be used for interpreting processes in nature.[18]

Another, this time nonmathematical, example of an analogy from nature for a technical application is the invention of Velcro. Here, the analogy does not model a process in nature. Holyoak and Thagard (1995, p. 198) have recounted the story: "In 1948, Georges de Mestral noticed that burdock burrs stuck to his dog's fur by virtue of tiny hooks. He figured out how to produce the same effect artificially, so that now shoes and many other objects can be fastened using burrlike hooks and clothlike loops. Velcro, which was originally a target analog for the burr source, has in turn become a source for further analogical designs and explanations, with targets drawn from medicine, biology, and chemistry. These new domains for analogical transfer include abdominal closure in surgery, epidermal structure, molecular bonding, antigen recognition, and hydrogen bonding." The interesting aspect of this story is that a technical application is developed drawing from nature and is subsequently again used as an analogue to account for processes in nature.[19]

The case of Velcro is very much an example of an analogy based on visual properties. There are others like it, the most notorious probably being Friedrich Kekulé's (1829–1896) dream of a snake biting its own tail, which in 1865 supposedly led him to the hypothesis that the carbon atoms in benzene are arranged in a ring. I phrase my reference to this example carefully, not because I want to imply that the story is untrue. The problem rather is that the story is based on an autobiographical report by Kekulé himself, years after the event, and there is much evidence from memory research that recollections are subsequently reconstructed. Also, Kekulé reconstructed his discovery story differently at different times (Shaffer 1994, p. 26). Moreover, if anything, I would expect that the reason Kekulé dreamed of the snake was related to his extended search for a solution concerning the chemical structure of benzene. However, there is no reason to deny that an unrelated event may trigger a scientist to find the missing link or to take a crucial conceptual step. Drawing an analogy between bacterial mutation and a slot machine helped Salvador Luria (1912–1991) to devise experiments in support of the thesis that phage-resistant bacteria come about because of gene mutations. Again, Holyoak and Thagard recount the story:

> In 1943, Salvador Luria was trying to find experimental support for his view that phage-resistant cultures of bacteria arise because of gene mutations, not because of action of the phage on the bacteria. (A phage is a viral organism that destroys bacteria.) None of his experiments worked, but at a faculty dance at Indiana University he happened to watch a colleague putting dimes into a slot machine. He realized that slot machines pay out money in a very uneven distribution, with most trials yielding nothing, some yielding small amounts, and rare trials providing jackpots. He then reasoned that if bacteria become resistant because of gene mutations, then the numbers of resistant bacteria in different bacterial cultures should vary like the expected returns from different kinds of slot machines. This reasoning led to an experiment and theoretical model, for which he was awarded a Nobel prize. (Holyoak and Thagard 1995, p. 188)

The visual analogies concerning Velcro, the benzene ring, and resistant bacterial cultures are instances where the analogy appears to have triggered a fairly sudden insight into a quite unrelated subject. All three examples are instances of very distant analogies; the source and the target belong to very different domains and the analogies are considered significant probably mostly because they led to what was later considered an important discovery.[20] But these analogies did not need to play a further role in the development of the target domain, other than trigger the idea for it. The technical intricacies of making Velcro, finding suitable material, and so on need not receive any significant input from thinking about burdock burrs, nor does one need to expect the statistics of slot machines to be the same as that of resistant bacterial cultures. And it was Kekulé's chemical knowledge and insight that enabled him to formulate and propose the ring shape in the structural formula for benzene, not any expertise regarding snakes. What is portrayed as a chance discovery is still, most of the time, the result of a very systematic search under systematically chosen research conditions.[21]

Another use of analogy in science is when an analogy is employed as part of an argument. Such analogies are a bit like thought experiments. Against the argument that the Earth does not move because a stone falling off a tower lands at the base of the tower, rather than some distance away that would correspond to the movement of the Earth, Galileo (1564–1642) formulated an analogy—namely, that a stone being dropped from the mast of a ship that is moving also lands at the bottom of the mast. This is in his *Dialogue Concerning the Two Chief Systems of the World* that was published in 1632. Gentner and Gentner 1983 was able to show that people's ability to solve problems regarding electric circuits depends on the analogy used: that of water flow and that of a moving crowd. Likening a resistor in an electrical circuit to a crowd trying to squeeze through a narrow gate, it appears plausible that with two "gates" in parallel, the resistance is obviously smaller, and with two "gates" in series, the resistance is higher.

Similarly, the voltage applied by a battery can be viewed in terms of the pressure difference produced by a reservoir (depending on the height of water in the reservoir).[22] However, neither the water nor the people analogy is suitable for *all* features of an electrical circuit. Gentner and Gentner (ibid.) showed that subjects who intuitively employ the moving-crowd model do better at problems concerning electrical resistance than those who use the water-flow model. In turn, the water-flow model grants more success with electrical circuits that involve batteries. Consequently, the analogy drawn in order to address a circuit problem has considerable impact on the reasoning in this type of physics problems. These are examples of analogies that can serve as arguments because they may incline one to favor one solution to a problem rather than another.

There are other uses of analogies, however, that are not just arguments but influence theory development. Gentner et al. (1997) have examined analogy as promoting creativity and conceptual change. Their claims about analogy in science are supported and illustrated by studying the case of Johannes Kepler (1571–1630). He used analogies extensively.[23] Often he even spelled out lines of reasoning that led him astray or turned out faulty. Kepler not only had to struggle with the transition from a Ptolemaic to a Copernican world system. Another assumption hard to overcome was the notion that the paths of planets must be explained in terms of perfect circular orbits (ibid., pp. 410ff.).

Moreover, Kepler tried to explain that the periods of the planets farther away from the Sun were longer, relative to those of the inner planets. The planets farther away from the Sun moved more slowly than those closer to it. So either the "moving soul" ("anima motrix") of the planets farther away was weaker, or the anima motrix was really located in the Sun ("vis motrix"). This meant that the objects farther away from the Sun were driven by it less vigorously than those nearer. Exploring this second possibility, Kepler drew an analogy to light. The light received from a light source per area also thins out with distance. The farther away, the lower the intensity of the light received (Gentner et al. 1997, pp. 414–415). This is an analogy to which Kepler returned repeatedly.

Kepler also studied the behavior of light in order to be better equipped for developing his analogy. A potential conceptual difficulty with a vis motrix located in the Sun is that it assumes action at a distance. Again, Kepler tried to justify this assumption on the basis of the analogy to light (ibid., p. 416). The aim of Gentner and her colleagues is to fit Kepler's use of the vis motrix analogy into their structure-mapping framework; they recognize the processes of highlighting, projection of the candidate inferences, re-representation and restructuring in Kepler's analogizing. Although there is no guarantee, they assemble a lot of evidence making it plausible that analogy drove conceptual change in the case of Kepler's work (ibid., pp. 438ff.), and the conceptual changes were indeed radical (ibid., pp. 443–445). Two main suggestions regarding creativity follow from this work of Gentner and her colleagues. One is that Kepler pursued his analogies extremely thoroughly; this is "the intensity of the alignment processes he carried out on his analogies once he had them in working memory" (ibid., p. 450). The other is the heterogeneity of Kepler's interests. Both of these factors are thought to have contributed to Kepler being capable of putting into place major conceptual changes. Kepler's case is a clear example of analogies being instrumental in the development of science.[24]

Finally, analogies are also sometimes used in teaching science. The main benefit of these analogies seems to be that they relate science to areas of familiarity that belong to the context of everyday life (Holyoak and Thagard 1995, pp. 199ff.). The risk involved in this strategy is that students can be misled by simplifying analogies. (For examples, see ibid., pp. 202f.) Teaching is not, however, directly related to modeling, even if models are sometimes taught, so I shall not follow this use of analogy any further.

Next I recapitulate the different uses of analogy in science introduced in this section and consider their relevance for modeling. Mathematical analogies can be relevant for modeling in those cases in which a model describes a phenomenon largely by means of mathematical equations. A mathematical analogy can help in formulating or understanding a model. Mathematical analogies that are used for data modeling, however, such as simulated annealing, have no direct impact on the model of the phenomena itself. There, the use of the analogy is technical and independent of the specific area of application of the technique. Similarly, the technical use of a nonmathematical scientific analogy, such as the development of Velcro, also has no direct relevance for modeling, because the goal of drawing such an analogy is *not* building a model of a phenomenon. Then there is the use of analogies in scientific discovery, as in Kekulé's dream of a snake biting its tail. Here, the analogy may provide the idea for the construction of a model, but it does not give *detailed* guidance for the construction of the model (no structure mapping; section 3.3). The analogy may function as a trigger, but it ultimately has no impact on the actual development and quality of a resulting model.

Finally, analogies used to develop an argument, as in the case of Kepler's analogy between light and vis motrix, are most relevant for scientific modeling. While the visual analogies discussed certainly have a momentary cognitive effect on theorizing, analogues that develop into an argument have a very extensive effect on the way in which scientists think about an issue. Note, however, that while an analogy can promote the development of a model more fruitfully or less so, it is nonetheless not responsible for the quality of the resulting model. The model has to live up to describing a phenomenon in question, whether or not the model construction has been inspired by or has relied upon an analogy. As emphasized before, the analogy is *not* the model. Whether a fertile analogy can be turned into a good model depends on what the modeler makes of the analogy. Some parts of the analogy need to be rejected, others exploited (negative, neutral, and positive analogy), and judgments have to be made about this in the course of modeling a phenomenon.

3.5 How to Study Analogies in Science

In the previous section I simply considered a range of examples of analogies occurring in science. Although the examples illustrate my points, they only give anecdotal evidence for the cognitive relevance of analogies in scientific thinking. In cognitive science, extensive work has been done on the subject of analogy, although employing a range of different approaches. Each approach has its methodological benefits and disadvantages. Correspondingly, it is often necessary to take the results of different approaches together, but keeping in mind the methodological limitations of individual approaches. In this section, I briefly sketch different approaches to the study of analogy, including some of their shortcomings. I do this to encourage drawing from the rich resources for the study of analogy, while being aware of the difficulties of applying results directly to the case of scientific reasoning.

A major source of evidence about analogy use are historical case studies and example cases. Such examples, some of them quoted throughout this chapter, can be found, for instance, in Turner 1955, Hesse 1966, North 1980, Gentner and Jeziorski 1989, Holyoak and Thagard 1995 (chapter 8), Gentner et al. 1997, Nersessian 2002, and Bailer-Jones and Bailer-Jones 2002. Diaries, notebooks, letters, and published articles can constitute the basis for such studies. The historical approach means that one can gain only limited access to the reasoning processes of scientists, be they analogical or not, because one is restricted to what scientists chose to write down. The information preserved or passed on to the reader is already filtered and selective.

The same holds, albeit perhaps to a lesser extent, for studies based on questionnaires or think-aloud protocols. The advantage of these methods is, however, that the analogy researcher can take a more active role in acquiring information. Another possibility is to rely on scientists' own reports about their cognitive processes. These may be found in autobiographies, in letters, or casual remarks on the side, but they can also be acquired by interviewing scientists. The problem with such evidence is that it may be biased due to the wish to present science—or oneself—in a particular way. Some things get misremembered, others forgotten. Kevin Dunbar's studies (1995 and 1997) of lab meetings in molecular biology labs show that sometimes scientists have already forgotten an analogy that they had fruitfully employed during a lab meeting, when interviewed a few days later. They only remember the result to which analogical reasoning may have led them. Autobiographical information, in a Nobel Prize speech or so, may often be colored so as to mythologize the discovery process. The case of Kekulé's snake is probably a good example for such mythologizing.

Other criticisms that have been put forward concerning the study of analogy in the cognitive psychology laboratory include the following (Dunbar 1995):

- In comparison to real discoveries, the social context is missing when fake discoveries are staged in the psychology lab. For instance, many scientists today work in groups rather than on their own, unlike the subjects of psychology experiments.
- In many experiments on analogy, the problems to be solved were not real scientific problems.
- The subjects of experiments on analogy use in science are frequently nonscientists. Alternatively, if the subjects are expert scientists, the problems to be solved are often not sufficiently scientific.
- The duration of lab experiments is often very short, unlike the duration of involvement in reasoning tasks in science.

Such considerations led Dunbar to the study of reasoning in scientific research in vivo—that is, not under special experimental conditions, but where science happens: in the lab. Dunbar (1995, 1997, and 1999) studied laboratory meetings in molecular biology and immunology laboratories, videotaping, audiotaping, and interviewing scientists "in action" to analyze their thinking. Not only is this a novel approach to the study of analogy, his work also yielded some unexpected results. First, the use of analogies in lab meetings is prolific. They are used for different purposes, such as formulating theories, designing experiments, and giving explanations to other scientists.

Depending on what the goal is, the type of analogies used varies (Dunbar 1995 and 1997). Distant analogies where the source domain differs significantly from the target domain (in this case, being non-biological) are quite rare. They may be used for explanation—for example, to instruct a novice. So long-distance analogies turned out not to be a driving force in discovery.

The use of "local" analogies stemming from a similar domain (for example, a previous experiment, or within the same organism) was extremely common. It usually occurred when technical difficulties arose. Regional analogies were also common and are those cases where entire systems of relationships are mapped and where the domains belong to different classes, but the same overall category (for example, different phage viruses—that is, another organism—but of the same group). This type of analogy serves the generation of hypotheses, the elaboration of theory, and the planning of new experiments. Generally, the frequency of analogy making in the different laboratories seems to have correlated with research progress made. For analogy making, it helps if the members of the research group do not have backgrounds too similar, and analogy making is often provoked in a group situation. Moreover, producing an analogy is often the immediate reaction to unexpected findings (Dunbar 1999).

Another possibility is to take analogy use in science simply as a specialized case of human analogy use more generally. In particular, parallels have been drawn between children's early cognitive development and scientific reasoning. Are children "scientists in the crib," as sometimes suggested (Gopnik, Meltzoff, and Kuhl 2000)? Is analogy use in science cultural or innate? Is there a continuum from baby to toddler to preschooler to Nobel Prize–winning scientist? Taking a developmental approach, sometimes the analogy use in children is studied (see Vosniadou 1989; Holyoak, Junn, and Billman 1984; Holyoak and Thagard 1995, chapter 4 and references therein; and Goswami 2001 for a comprehensive and up-to-date review of analogical reasoning in children). Evidence for early analogy use is found, getting gradually less similarity oriented, and the question then arises whether these findings illuminate patterns of scientific reasoning.

Rather than comparing scientists' analogical reasoning to children, they are also often compared to computer programs simulating the reasoning process. The assumption underlying such approaches is that we can learn about how humans reason ("how the brain works") from the way reasoning processes can be implemented on a computer. Again, an analogy is exploited in this very project: that between the real and the simulated process. In other words, the computational implementation is used as an analogue to the mental processes involved in analogical reasoning. This analogy is thought necessary because it is impossible to "inspect" the mind directly (Holyoak and Thagard 1995, p. 238). The test for such a computer implementation of analogical reasoning is whether the computer comes up with the same kind of inferences and reasoning errors as found in human subjects studied in psychology experiments. Although Holyoak and Thagard (1997, p. 42) have acknowledged that computational models of analogical reasoning are surpassed by human use of analogy, they still expect that this "continued development of more sophisticated computational models will lead to deeper understanding of the analogical mind" (ibid., p. 43).[25] Another criticism that has been levied against some such approaches is that the apparent success of these computer programs tells us not much about analogical reasoning, because, to produce the desired results, their output is a direct function of their input. To put it differently: so much information about the analogues is part of the input that the resulting inferences are no real creative achievement. Finally, as in many psychology experiments, the cases of analogical reasoning implemented and studied computationally are often far less complex and far less extended than those in real science.[26]

There can be absolutely no doubt that research on the role of analogy in human reasoning, and on analogy use in science, is prolific. The sheer range of different studies seems to count as an argument that there is something to the cognitive force of analogy—that there is something to the claim that the use of analogy is a central and indispensable strategy for interpreting the world. Insofar as this is correct, the use of analogy supports scientists' modeling efforts. Having highlighted the methodological limitations of a range of different types of studies of analogy use, however, also indicates that it is never easy to apply the results of such studies directly to very specific and very complex tasks, such as scientific modeling.

3.6 Conclusion

In this chapter I considered the role of analogy because analogies are often quite central to scientific modeling, to the extent that it has sometimes been claimed that analogies *are* models. In section 3.1, I reviewed how scientists commented on the use of analogy in the nineteenth and early twentieth centuries, at a time when scientific modeling, analogy use, and the mathematization of physics advanced significantly. The themes emerging in this context were the use of analogy for knowledge generation and the deepening of domain knowledge. Analogizing was highlighted as a method of hypothesis formulation and discovery, not to talk about illustration and instruction. Analogies were seen as useful in linking different areas of investigation, helping to interpret mathematical descriptions of new domains. These fairly intuitive views of scientists later reappeared in the work of some philosophers of science and in Achinstein's claim that models have an explanatory function (section 3.2).

I presented the formal analysis of what an analogy is and does in section 3.3. A major distinction is that between material and formal analogies, or whether a source and a target domain have attributes or relations—in Gentner's terms, structure—in common (or both). In section 3.4, I introduced and discussed a number of well-known analogies used in the sciences, illustrating that there are a number of different uses of analogy in the process of science: emphasizing links between unrelated domains, inspiring technical applications, provoking sudden insights, strengthening an argument, promoting theory development. As analogy has attracted attention from a number of different disciplines—among them philosophy, history of science, cognitive psychology, and computer science—I outlined the pros and cons of different types of analogy study and how they can contribute to our understanding of the role of analogy in science (section 3.5). The short story is that no method is perfect, but complementing each other, the different research strategies result in a fairly convincing picture of the importance of analogy use in science.

So what are the repercussions of the findings from analogy study regarding the study of scientific models? It is clear that analogies can be the foundation of a model—that is, lead to its conception and further its development (for example, Kepler's analogy between light and the vis motrix). However, models, as they mature, also specify those parts of the description of a phenomenon that are unlike the source domain. In Hesse's terms, the negative analogy is not part of the model. It will often take some time to work out and research which neutral analogies have to be discounted as negative analogies. Just as analogies can be employed during different stages of the scientific process, models can also be tentative or advanced and elaborated, but in any case the analogy is not the model. An analogy is a relation, whereas a model is a (partial) description of a phenomenon. One way of viewing the relationship between model and phenomenon is as one of analogy, loosely in the tradition of Campbell, although this is now not a common way of talking about analogy in science. Following the structure-mapping approach, this would then imply that similar (or identical) relationships hold in the phenomenon as in the model.[27]

From the point of view of assessing a model, whether there exists an analogy on which the model has been based is ultimately not decisive, even if, in practical terms, the analogy has been instrumental for the model development and continues to be cognitively instrumental for handling the model. A model may display distinct traces of having originated in a certain analogy, but as a model it is judged regarding the phenomenon of which it is a model. Is the model a good description, does it represent the phenomenon well? (For more on what this means, see chapter 8.) However important analogy is as a cognitive tool and strategy in science, there is no reason to think of it as the only cognitive strategy by which to arrive at a model (Bailer-Jones 2000), nor as the only route toward grasping that model.

Notes

1. Hesse ([1963] 1966, for example, pp. 3ff.) discusses Campbell's ([1920] 1957) views on models, despite the fact that Campbell never mentions the term "model" and only ever talks about analogy. Campbell's view on the role of analogy in science is discussed in section 3.2.

2. I restrict my discussion to *scientists* analyzing the use of analogy in science. John Stuart Mill (1806–1873), for instance, was concerned with the method of science and even discusses analogy in his *A System of Logic*, but he

is not a scientist. John North (1980, p. 123) has commented: "Both Bacon and Mill paid attention to analogy, in their account of induction, and Mill's passage on the subject was especially influential in philosophical circles. I am sure, nevertheless, that neither Mill nor Bacon had very much influence on the analogical *techniques* of scientists."

3. For a detailed discussion of Maxwell's use and concept of analogy, see Turner 1955 and Nersessian 2002. See also Maxwell (1856) 1890.

4. North (1980) highlights Thomson's major influence on Maxwell's use of analogy. He wrote: "To Thomson belongs the credit for drawing the attention of physicists to the *power* of analogy. He did so, not by writing a logic of analogy, but by developing a notable *example*" (ibid., p. 123). For a list of the analogical concepts in this example, see ibid., p. 124. For more analogies used and introduced by Thomson, see ibid., pp. 124–125.

5. In what sense this type of knowledge is more profound than other types would require spelling out.

6. Mellor (1968) further claims that were it not for Campbell's strict theory-observation distinction, his account would not differ significantly from Duhem's.

7. This interpretation had some repercussions in philosophy of science. Both Hesse (1953, p. 201; see also Hesse 1967) and Harré (1970, p. 35) also propose (besides analogies *between* the theoretical treatments of phenomena) that scientific models are to be viewed as analogues to the aspects of the real world that are their subject. Rom Harré has called this "[a] behavioural analogy between the behaviour of the analogue of the real productive process and the behaviour of the real productive process itself" (Harré 1988, p. 127).

8. Not everybody agrees. Carl Hempel (1965, p. 440), for instance, states categorically: "[A]ll references to analogies or analogical models can be dispensed with in the systematic statement of scientific explanations."

9. There is obviously no reason to claim that analogy is the only creative strategy employed in scientific reasoning. For an identification of other strategies found in in vivo research carried out in a laboratory setting, see Dunbar 1997.

10. Toward the end of the essay, Hesse (1966, p. 96) also considers the case where some of the properties of two analogues are observables, whereas at least one may be a theoretical term, in which case the analogy can help to define the theoretical term. Claiming that the properties relevant for a material analogy are observable is subject to all the usual problems with the theoretical-observable distinction, and Hesse clearly does not rule out taking into account theoretical properties in principle. Correspondingly, I suspect there is little point being strict about only admitting observable properties. See also Hesse's (ibid., p. 15) reference to the problems with the very term "observable."

11. The analogy can have the status of an assertion about the relationship between the state and its citizens that may be intended to be persuasive or normative.

12. North (1980, pp. 134ff.) also sheds doubt on the possibility of a clear-cut distinction between material (which he calls "substantial") and formal analogy; it only applies in special cases. He considers this distinction as "a by-product of a traditional phase of analysis of the logic of analogy which should have ended at the time of Thomson and Maxwell, if not before" (ibid., p. 135)—the argument being that during this phase undue attention was paid to analogies between *things* (or *terms*). It is not the kind of distinction that has much to do "with the problem set by those who, in the natural sciences, have *advanced* their knowledge by analogical techniques that they have seldom tried to justify" (ibid., p. 136). Analogy has the purpose of being a tool of exploration, no matter what its exact logical and epistemological status.

13. This approach has subsequently been computer-implemented (Falkenhainer, Forbus, and Gentner 1989; see also section 3.5).

14. Gentner (1982, for example, p. 109) views this characteristic of analogy as specific to science, in that it deals with comparisons of systems, rather than just comparisons of objects.

15. Gentner and her collaborators also extensively studied the relationship between analogy and similarity. According to them, drawing an analogy and making out a similarity are the same psychological processes, and this is why they extend the structure-mapping approach to similarity. For an elaboration of this claim, see Gentner and Markman 1997.

16. Gentner and Jeziorski (1989) have compared analogy use in science as practiced by alchemists as well as by scientists such as Robert Boyle (1627–1691) and by Sadi Carnot (1796–1832). They conclude that there are different styles of analogical reasoning and that rigorous use of analogies in science has only developed over time and is not universal. Gentner and Jeziorski (ibid., p. 320) conclude that "the rules of analogical soundness are not innate. Despite the seeming inevitability of the analogical precepts we now use, they are not a necessary part of natural logic."

17. In chapter 2, section 2.2, I highlighted that models in the nineteenth century were models that were physically built, whereas today we would also refer to some theoretical constructs as models—for example, to the analogies Maxwell used (the "vortex model").

18. Given that the analogy serves to develop a data analysis technique, the analogy is instrumental in developing something that, as a technique, is "application-neutral" (Bailer-Jones and Bailer-Jones 2002, pp.159ff.). Such an application-neutral technique can, however, still play a major role in modeling *natural* processes. The difference is that the analogy is not formulated with regard to a process in nature, unlike most other cases of the use of analogy in science that are discussed here.

19. In the case of simulated annealing, it was really drawing from a *description* of nature.

20. According to Dunbar 1997, the majority of analogies that are instrumental in discovery and scientific progress are *not* from a very different do-

main, but from one very similar to the one in which a question arises; see section 3.5.

21. Compare Dunbar's 1999 findings concerning "chance discovery."

22. Two reservoirs in parallel—that is, on the same level—both have the same height of water, thus the pressure they produce is no higher than it would be with only one reservoir. If the reservoirs are "in series," however—that is, one on top of the other—then the height of the water is doubled and they produce twice the original pressure, in the same way as two identical batteries in series produce double the voltage of one.

23. According to Gentner and her colleagues (1997, p. 406): "Kepler was a prolific analogizer. Not only in his books but also in his journals and letters, he used analogies constantly. In some cases the analogies seem playful. In other cases, analogizing is integral to his theorizing."

24. Sometimes an analogy may be pursued with considerable ferocity and yet not be blessed with lasting success. Following the analogy between light and sound, Isaac Newton drew analogies between the seven tones in music and the seven colors he perceived in the optical spectrum. He hoped in vain that the musical scale would prove a useful conceptual tool for investigating the phenomena of light and color (North 1980, pp. 116–121).

25. One difficulty with this approach is that there are, of course, a number of substantially different ways of implementing analogical reasoning successfully, and it is then not at all obvious which computational approach is supposed to present the best analogue to human analogical reasoning. For an accessible introduction to such approaches, see Holyoak and Thagard 1995, chapter 10. Well-known examples are the Structure Mapping Engine (SME) (Falkenhainer, Forbus, and Gentner 1989), Analogical Mapping by Constraint Satisfaction (AMCE) (Holyoak and Thagard 1989a and 1989b), or Copycat (Hofstadter 1995).

26. This is so with the possible exception of Gentner, who attempted to formalize and computationalize scientific reasoning processes (for example, in Gentner et al. 1997).

27. Compare this with the notion that model and phenomenon are isomorphic, as sometimes put forward in the context of the Semantic View (see chapter 6, section 6.1).

References

Achinstein, P. 1968. *Concepts of Science.* Baltimore, Maryland: John Hopkins University Press.

Bailer-Jones, D. M. 2000. Scientific models as metaphors. In F. Hallyn, ed., *Metaphor and Analogy in the Sciences.* Dordrecht: Kluwer Academic Publishers, pp. 181–198.

———, and C. A. L. Bailer-Jones. 2002. Modelling data: Analogies in neural networks, simulated annealing, and genetic algorithms. In L. Magnani and N. Nersessian, eds., *Model-Based Reasoning: Science, Technology, Values.* New York: Kluwer Academic/Plenum Publishers, pp. 147–165.

Campbell, N. R. [1920] 1957. *Foundations of Science* (formerly titled *Physics, the Elements*). New York: Dover Publications.

Clement, C. A., and D. Gentner. 1991. Systematicity as a selection constraint in analogical mapping. *Cognitive Science* 15: 89–132.

Duhem, P. [1914] 1954. *The Aim and Structure of Physical Theory.* Translated from the French 2nd edition. Princeton, New Jersey: Princeton University Press.

Dunbar, K. 1995. How scientists really reason: Scientific reasoning in real-world laboratories. In R. J. Sternberg and J. Davidson, eds., *Mechanisms of Insight.* Cambridge: MIT Press, pp. 365–395.

———. 1997. How scientists think: On-line creativity and conceptual change in science. In T. B. Ward, S. M. Smith, and S. Vaid, eds., *Creative Thought: An Investigation of Conceptual Structures and Processes.* Washington, D.C.: APA Press, pp. 461–493.

———. 1999. How scientists build models: In vivo science as a window to the scientific mind. In L. Magnani, N. Nersessian, and P. Thagard, eds., *Model-Based Reasoning in Scientific Discovery.* New York: Plenum Press, pp. 89–98.

Falkenhainer, B., K. D. Forbus, and D. Gentner. 1989. The structure-mapping engine: An algorithm and examples. *Artificial Intelligence* 41: pp. 1–63.

Gentner, D. 1982. Are scientific analogies metaphors? In D. S. Miall, ed., *Metaphor: Problems and Perspectives.* Sussex: The Harvester Press, pp. 106–132.

———. 1983. Structure mapping: A theoretical framework for analogy. *Cognitive Science* 7: 155–170.

———, and A. B. Markman. 1997. Structure mapping in analogy and similarity. *American Psychologist* 52: 45–56.

———, and D. Gentner. 1983. Flowing waters and teeming crowds: Mental models of electricity. In D. Gentner and A. L. Stevens, eds., *Mental Models.* Hillsdale, New Jersey: Lawrence Erlbaum Associates, pp. 99–129.

———, and M. Jeziorski. 1989. Historical shifts in the use of analogy in science. In B. Gholson, W. R. Shadish Jr., R. A. Beimeyer, A. Houts, eds., *The Psychology of Science: Contributions to Metascience.* New York: Cambridge University Press, pp. 296–325.

———, S. Brem, R. W. Ferguson, P. Wolff, A. B. Markman, and K. Forbus. 1997. Analogy and creativity in the works of Johannes Kepler. In T. B. Ward, S. M. Smith, and J. Vaid, eds., *Creative Thought: An Investigation of Conceptual Structures and Processes.* Washington, D.C.: American Psychological Association, pp. 403–459.

Gopnik, A., A. N. Meltzoff, and P. K. Kuhl. 2000. *The Scientist in the Crib: What Early Learning Tells Us about the Mind.* New York: Harper Collins.

Goswami, U. 2001. Analogical reasoning in children. In D. Gentner, K. J. Holyoak, and B. N. Kokinov, eds., *The Analogical Mind: Perspectives from Cognitive Science.* Cambridge: MIT Press, pp. 437–470.

Harré, R. 1960. Metaphor, model, and mechanism. *Proceedings of the Aristotelian Society* 60: 101–122.
———. 1970. *The Principles of Scientific Thinking.* London: Macmillan.
———. 1988. Where models and analogies really count. *International Studies in the Philosophy of Science* 2: 118–133.
Hempel, C. 1965. *Aspects of Scientific Explanation, and Other Essays in the Philosophy of Science.* New York: Free Press.
Herschel, J. F. W. [1830] 1987. *A Preliminary Discourse on the Study of Natural Philosophy.* Chicago: University of Chicago Press.
Hesse, M. 1953. Models in physics. *British Journal for the Philosophy of Science* 4: 198–214.
———. 1966 [1963]. *Models and Analogies in Science.* Notre Dame, Indiana: University of Notre Dame Press.
———. 1967. Models and analogy in science. In P. Edwards, ed., *The Encyclopedia of Philosophy.* New York: Macmillan, pp. 354–359.
Holyoak, K., and P. Thagard. 1989a. A computational model of analogical problem solving. In S. Vosniadou and A. Ortony, eds., *Similarity and Analogical Reasoning.* Cambridge: Cambridge University Press, pp. 242–266.
———. 1989b. Analogical mapping by constraint satisfaction. *Cognitive Science* 13: 295–355.
———. 1995. *Mental Leaps: Analogy in Creative Thought.* Cambridge: MIT Press.
———. 1997. The analogical mind. *American Psychologist* 52: 35–44.
Holyoak, K. J., E. N. Junn, and D. Billman. 1984. Development of analogical problem-solving skill. *Child Development* 55: 2042–2055.
Hofstadter, D. R. 1995. *Fluid Concepts and Creative Analogies: Computer Models of the Fundamental Mechanisms of Thought.* New York: Basic Books.
Markman, A. B. 1997. Constraints on analogical inference. *Cognitive Science* 21: 373–418.
Maxwell, J. C. [1855] 1890. On Faraday's lines of force. In W. D. Niven, ed., *The Scientific Papers of James Clerk Maxwell.* Vol. 1. Cambridge: Cambridge University Press, pp. 155–229.
———. [1856] 1990. Essay for the Apostles on "Analogies in Nature." In P. M. Harman, ed., *The Scientific Letters and Papers of James Clerk Maxwell.* Vol. 1. Cambridge: Cambridge University Press, pp. 376–383.
———. [1870] 1890. Address to the mathematical and physical sections of the British Association. In W. D. Niven, ed., *The Scientific Papers of James Clerk Maxwell.* Vol. 2. Cambridge: Cambridge University Press, pp. 215–229.
Mellor, D. H. 1968. Models and analogies in science: Duhem *versus* Campbell? *Isis* 59: 282–290.
Nagel, E. [1960] 1979. *The Structure of Science.* Indianapolis, Indiana: Hackett Publishing Company.
Nersessian, N. J. 2002. Maxwell and "the method of physical analogy": Model-based reasoning, generic abstraction, and conceptual change. In

D. Malamant, ed., *Essays in the History and Philosophy of Science and Mathematics.* Lasalle, Illinois: Open Court, pp. 129–166.

North, J. D. 1980. Science and analogy. In M. D. Grmek, R. S. Cohen, and G. Cimino, eds., *On Scientific Discovery.* Boston Studies in the Philosophy of Science. Dordrecht: D. Reidel Publishing Company, pp. 115–140.

Shaffer, S. 1994. Making up discovery. In M. Boden, ed., *Dimensions of Creativity.* Cambridge: MIT Press, pp. 13–51.

Turner, J. 1955. Maxwell on the method of physical analogy. *British Journal for the Philosophy of Science* 6: 226–238.

Vosniadou, S. 1989. Analogical reasoning as a mechanism in knowledge acquisition: A developmental perspective. In S. Vosniadou and A. Ortony, eds., *Similarity and Analogical Reasoning.* New York: Cambridge University Press, pp. 413–437.

THEORIES 4

In the work of Faraday, Kelvin, and Maxwell, models were central tools for the development of scientific accounts. These models were predominantly mechanical models, and they frequently guided the mathematical treatment of increasingly less mechanical phenomena, such as electromagnetism (see chapter 2). Mathematical treatment of phenomena, in turn, is one factor that furthers the emergence of theoretical concepts. Although mechanical approaches maintain the appearance of familiarity so that it would seem as if the mechanically explained process in question is directly observable, this is blatantly not the case when the approach becomes more mathematical. Indeed, the mathematization and abstractness of physics went to new heights—for example, with the non-Euclidean geometry employed in Einstein's Theory of Relativity or with quantum theory. Modern physics seemed to require that physicists focus on interpreting empirical phenomena in abstract and theoretical terms, marked by the extensive use of specialized mathematical methods in physics, worlds apart from the mechanical models that could either be physically built or at least be observed before the mind's eye.

The goal of this chapter is to analyze why, despite their practical importance in nineteenth-century science, models did not receive any (positive) attention during the first half of the twentieth century. This is very much a negative piece of history of philosophy of science regarding models, emphasizing what did not get said about models. Yet the almost total disregard of the role of models for a fifty-year period needs explaining. For this reason, this chapter talks much about the role attributed to theories. The received opinion was that good theories rendered models theoretically and practically redundant. As a hangover from the early days of philosophy of science, even now the term "model" very often still implies that a chosen description is in a merely preliminary version, to be confirmed later. "Theory" has the connotation of being well established and well abstracted—that is, free from the little failings and inaccuracies of models.

According to this common perception, models are mainly of temporary benefit; they come and go, while theories last. Models are thought to be accessible at the expense of the correctness, generality, and precision of theories. As a consequence, it requires less commitment or conviction to call something a model simply because it is "only a model," still awaiting correction and improvement. The underlying expectation is always that in the longer term the model will be replaced once a definitive account has emerged in the form of a theory. However, this received attitude toward models not only conceals the benefits and advantages scientists derive from the use of models, but it leaves important philosophical issues ignored that, after a long period of disinterest and even ridicule, led to a philosophical reevaluation of scientific models. The historical lesson of this chapter meets its philosophical response in chapter 6, where I argue how and why the balance between theories and models needs to be redressed.

In section 4.1, I begin with the successful application of mathematical descriptions that supported theory dominance, picking up a thread from chapter 2, section 2.1. Theory taking axiomatic form and being reconstructed according to the "hypothetico-deductive method" is introduced in section 4.2. Pierre Duhem, who made appearances in chapters 2 and 3, has reached notoriety for playing down models as a tool for lesser minds, but he is also a proponent of the hypothetico-deductive approach to theory. His influential views on the matter are reviewed in section 4.3. Formulating theories also involves postulating theoretical entities. Section 4.4 introduces the operationalist and Logical Empiricist responses to the problem of linking theoretical postulates to observational evidence.[1] The Logical Empiricist emphasis on the context of justification, rather than the context of discovery, is a further building block in accounting for the disregard of models under the influence of Logical Empiricism (section 4.5). Section 4.6 shows how early considerations concerning models come about in response to the philosophical issues outlined in section 4.4. In section 4.7, I present early positions involving models formulated by British philosophers in the 1950s. These early discussions of models raise many of the major issues that determine contemporary philosophical discussion (section 4.8). These issues are also the topics of chapters 5, 6, and 8.

4.1 Mathematical Predictions

As the nineteenth century drew to a close, employing theories that heavily relied on a mathematical apparatus had turned out to be a promising strategy in generating scientific knowledge and seemed to compensate scientists for the lack of direct observability in the generation of scientific knowledge (Gower 1997, p. 131). There could be no doubt that experimentation was only one part of the strategies for acquiring scientific knowledge. Endorsement for theoretical, mathematical approaches came whenever mathematically derived consequences of theories could be validated by experiment retrospectively. A famous classical example for this is the proposal of Augustin Jean Fresnel (1788–1827) that optical diffraction could be explained assuming that light had wave properties. He submitted this proposal for the Grand Prix of the Académie des Sciences of 1819. Siméon-Denis Poisson (1781–1840) challenged it, showing that wave properties would allow interference to occur, and interference would cause a bright spot to occur in the center of the shadow of a disk-shaped object when light shines at the screen from a point source. This bright point at first seemed an absurd consequence and counted as an argument against Fresnel's proposal. François Arago (1786–1853) asked, however, to have Poisson's prediction tested, and when the bright spot could indeed be observed, confirming Fresnel's proposal, this caused considerable amazement (Harman 1982, pp. 21–24).

A similar case is Maxwell's mathematically developed theory of electromagnetic waves. Only about twenty years after its publication, Heinrich Hertz (1857–1894), in 1888, was able to confirm it experimentally. He produced electromagnetic waves, measured their wavelength, and showed that their finite velocity equaled the velocity of light. Although it is hard to argue why mathematical description and derivations from it should play such a dominant role in the formation of scientific knowledge, successful predictions like these fostered the faith in the value of mathematical descriptions and derivations of physical phenomena.

4.2 Hypothetico-deductive Reconstruction of Theory

The close link forged between theory and mathematics led to theory being fashioned in the image of mathematics. To achieve this high goal, the theory itself sometimes had to be reconstructed retrospectively to approach axiomatic form. Consequently, it was often only some time after the messy discovery process that a tidy and apparently exception-less theory evolved. It is only with hindsight that theory can be presented in the image of mathematics, skinned of the historical details, tentativeness, and accidents of the discovery process.

The modern, now classical form for a systematic presentation of theory is the axiomatic form. The blueprint for this is the complete axiomatization of Euclidean geometry by David Hilbert (1862–1943). When theories are axiomatized, the axioms make no specific reference to the empirical foundation of the theory. On the contrary, what are to be the empirical consequences of the theory are *deduced* from the axioms in the expectation that the deduced will be in agreement with the observed empirical findings. This is why the axioms may be seen to have the status of hypotheses, thus the "hypothetico-deductive method." Heinrich Hertz, for instance, chose this axiomatic method when he sought a formulation of Maxwell's theory that was, from his perspective, more straightforward than Maxwell's own. Maxwell's formulation is shaped by the discovery process and betrays its roots in outdated precursor theories—for example the vortex model—so sometimes relies on concepts that may subsequently appear redundant (Fölsing 1997, p. 452).

In Hertz's approach, axioms replaced theoretical postulates that may in practice not always have appeared very satisfactory. Hertz undertook this project in his article *Über die Grundgleichungen der Elektrodynamik für ruhende Körper* (Hertz [1890] 1984). He may have taken his lead from the success of the axiomatic method in thermodynamics that deduces everything from two postulates: the conservation of energy and the increase of entropy (Fölsing 1997, p. 453). The pursuit of the axiomatic project also revealed Hertz's bias toward a philosophical program of "removing everything superfluous from theory" and "reducing it to its simplest form," as Hertz expressed it in his diary (ibid., pp. 450–451). Hertz was a gifted and highly successful experimenter, paying due tribute to the ongoing need for experimentation, but he also emphasized theory as a central means of expressing physical insight. In an August 10, 1889, letter to Oliver Heaviside, he wrote: "Theory goes much further than the experiments, for the experiments hardly come to tell in a whispering voice what theory tells in clear and loud sentences. But I think in due time there will come from experiment many new things which are not now in the theory, and I have even now complaint against theory, which I think cannot be overcome until further experimental help" (cited in O'Hara and Pricha 1987, p. 68). It is a matter of choice and practicality for scientific practice that Hertz presented Maxwell's theory in an axiomatic form that is removed from the historical process of theory development. Theory is then deliberately reconstructed to fit the axiomatic form. The case of Hertz's work on Maxwell's theory shows that a hypothetico-deductive presentation of a theory relies significantly on reconstruction after the discovery process.

4.3 Duhem on Theories and Models

To philosophically minded physicists, the hypothetico-deductive method had considerable attraction. The work of Duhem shows how adherence to this theory ideal is combined with the rejection of models as a desirable tool of science. We encountered Duhem already in chapter 3 (section 3.1) considering his views on analogy, but more influential is his work on theories. Duhem, a physicist, philosopher, and historian of science, not only favored a hypothetico-deductive account of theories, but he also expressed a most distinct preference for theories over models. His views both on theories and on models exerted considerable influence in philosophy of science. Duhem ([1914] 1954, p. 19) presents theory as a deductively closed system: "[A physical theory] is a system of mathematical propositions, deduced from a small number of principles, which aim to represent as simply, as completely, and as exactly as possible a set of experimental laws."

This definition is very much in accordance with what Hertz practiced when he presented Maxwell's electrodynamic theory axiomatically. Duhem goes on to spell out the nonmetaphysical status of theories as hypothetical, with the principles of theories not claiming "to state real relations among the real properties of bodies" (ibid., p. 20). How closely theory is conceived in the image of mathematics is evident from the following quotation: "The diverse principles or hypotheses of a theory are combined together according to the rules of mathematical analysis. The requirements of algebraic logic are the only ones which the theorist has to satisfy in the course of this development. The magnitudes on which his calculations bear are not claimed to be physical realities, and the principles he employs in his deductions are not given as stating real relations among those realities; therefore it matters little whether the operations he performs do or do not correspond to real or conceivable physical transformations. All that one has the right to demand of him is that his syllogism be valid and his calculations accurate" (ibid., p. 20).

The point where theory parts with pure mathematics is the requirement that the consequences of the deductions and derivations are in agreement with the experimental laws. He wrote: "*Agreement with experiment is the sole criterion of truth for a physical theory*" (ibid., p. 21; italics in original). However, the nature of this agreement and how agreement is established is left vague at this point, something that became a major concern in the first decades of the twentieth century. Importantly for twentieth-century philosophy of science, Duhem took a firm stance regarding scientific models. While at his heart Duhem seems to have dismissed models, he was historically well informed of their use in the nineteenth century. He therefore gave a spirited account of models that insightfully characterizes, even if in the form of caricature, their uses.

So, although Duhem played a major part in the neglect and rejection of models in the philosophy of the first half of the twentieth century, he also inadvertently inspired their study later on. The broad picture that Duhem proposes of the relation between theory and models is thus: "The descriptive part has developed on its own by proper and autonomous methods of theoretical physics; the explanatory part has come to this fully formed organism and attached itself to it like a parasite" (ibid., p. 32). "The descriptive part" refers to theory, while "the explanatory part" refers to models. According to this, models feature as no more than appendices to theories; they have no crucial role and are redundant as far as any relevant aspect of scientific development is concerned.[2] Polemically distinguishing between the abstract and narrow (French) mind and the ample and weak—but imaginative—(English) mind, Duhem considers hypothetico-deductive theories, in contrast to mechanical models, as the carriers of scientific progress. He wrote: "[O]nly abstract and general principles can guide the mind in unknown regions and suggest to it the solutions of unforeseen difficulties" (ibid., p. 93).

Despite this preconceived opinion—or prejudice—Duhem does a good deal of justice to scientific models. Not only abstract minds can be vigorous; vigorous ample minds are able to have a "vision": "There are some minds that have a wonderful aptitude for holding in their imagination a complicated collection of disparate objects; they envisage it in a single view without needing to attend myopically first to one object, then to another; and yet this view is not vague and confused, but exact and minute, with each detail clearly perceived in its place and relative importance" (ibid., p. 56). Although there may be some benefits to approaching nature with an ample mind, physical mechanical models (cf. chapter 2, section 2.2) do not fare well in Duhem's polemics portraying the English use of models: "Here is a book (O. Lodge, op. cit.) intended to expound the modern theories of electricity and to expound a new theory. In it there are nothing but strings which move around pulleys, which roll around drums, which go through pearl beads, which carry weights," and so it goes until it famously culminates in "We thought we were entering the tranquil and neatly ordered abode of reason, but we find ourselves in a factory" (ibid., pp. 70–71).

With this caricature, Duhem did his best to reduce the concept of modeling to that of building mechanical machines. He characterized models thus as outdated, without considering a possible expansion of the concept that would adjust to the mathematization of physics, as has later been attempted (Hesse 1953). Once an extended concept of mechanism is contemplated (see chapter 2, section 2.4), Duhem's description of the English mind loses some of its sting. He comments: "Understanding a physical phenomenon is, therefore, for the physicist of the English school, the same thing as designing a model imitating the phenomenon; whence the nature of material things is to be understood by imagining a mechanism whose performance will represent and simulate the properties of the bodies" (Duhem [1914] 1954, p. 72). Allowing for models more abstract than the potentially inferior mechanical machines, many modern scientists would suddenly appear to have something of an Englishman in them whose object it is "to create a visible and palpable image of the abstract laws that [the Englishman's] mind cannot grasp without the aid of this model" (ibid., p. 74). Thus Duhem stands for the dominance of the mathematical, deductive method in scientific theorizing. Taken out of his narrow context, however, he sounds rather more farsighted with his talk about "creating an image" in the face of abstract laws (cf. chapter 5).

4.4 Theoretical Postulates

One consequence of the hypothetico-deductive method is that some of the entities featuring in the axioms of the theory are postulated without explicit link to the empirical data. This is an issue that dominated the discussion surrounding theories during the 1920s and 1930s, and it is an issue that models were able to address in the 1950s.

Electromagnetic waves are a striking example where the mathematical exploration of a physical theory leads to the postulation of an entity that has not been observed directly. Even when observational evidence could be produced, it was hardly the case that the waves were out there for everybody to see. Instead, some properties of the wave—for example, interference—were examined that then allowed the observer to infer some characteristics that indicated that the wave was indeed a wave of a certain kind. In a similar vein, Rudolf Carnap (1891–1970) said about the electrical field: "And the predicate 'an electrical field of such and such an amount' is not observable to anybody, because, although we know how to test a full sentence of this predicate, we cannot do it directly, i.e. by a few observations" (Carnap 1936, p. 455).

Empirical data that provide evidence for an entity are not the same as direct observation of an entity. The track an electron leaves in a cloud chamber is probably the prototypical example for such indirect observational evidence for a theoretically postulated entity, the electron. Not only is the observation of something like an electrical field indirect, its assumed existence also depends on its occurrence in a theory. If a theory postulates that a certain entity exists, then the question arises how this postulate is linked to the observational data. As part of the empiricist program, the aim was to show that even theoretical postulates—defined negatively as not being part of the observational vocabulary (cf. Danto and Morgenbesser 1960, p. 25; and Hesse 1970, pp. 36 and 73)—are ultimately based on experiences. Only that which is based on sensory experiences counted as constituting knowledge. The question therefore was how theoretical postulates relate to sensory experiences—that is, that which can be observed. This general outlook, positivism, has, as Mary Hesse has noted, a number of roots. Besides Auguste Comte (1798–1857), who is regarded as the founder of positivism, these roots stretch from Ernst Mach (1838–1916) for the German-speaking world to David Hume (1711–1776) for the British. Hesse wrote (1954, pp. 92–93): "The word 'positivism' has now come to denote so many different points of view that it has almost lost its usefulness, but it can be taken at least to denote a rejection of most of traditional metaphysics, and its replacement by emphasis on 'that which is laid down' (*positum*), that is, on the immediate data of sense experience as described by the empirical philosophers, particularly by Hume."

Carnap described the issue of relating theoretical postulates to observational experience in terms of language—that is, in terms of reducing (theoretical or abstract) terms of a scientific language to sense-data (or observational) terms (without himself claiming that such a reduction is possible in any full sense). He commented: "One of the fundamental theses of *positivism* may perhaps be formulated in this way: every term of the whole language L of science is reducible to what we may call sense-data terms or perception terms" (Carnap 1936, pp. 463–464).

Proposing solutions to the problem of linking theoretical postulates with empirical data became a major issue in thinking about science in the 1920s and 1930s. Next I discuss a few important examples. Operationalism is one approach that addresses the problem of theoretical terms or "physical quantities" (Eddington [1923] 1975) in which an explicit definition of theoretical terms via what is observable is sought. This definition is expected to be achievable by referring to the experimental procedures and their observational results. Certainly in the 1920s, operationalism was a popular philosophical outlook among physicists. Perhaps the best known but by no means the only exposition of it is that of the American physicist Percy Bridgman (1882–1961) in *The Logic of Modern Physics* (Bridgman [1927] 1961). Arthur Eddington (1882–1944), for instance, dedicates the introduction to his *The Mathematical Theory of Relativity* to laying out this position: "To find out any physical quantity we perform certain practical operations followed by calculations; the operations are called experiments or observations according as the conditions are more or less closely under our control. The physical quantity so discovered is primarily the result of the operations and calculations; it is, so to speak, *a manufactured article*—manufactured by our operations" (Eddington [1923] 1975, p. 1).

The obvious difficulty with this approach is that experimental procedures for establishing certain theoretical entities can be varied and that they may change over time (cf. Hempel [1954] 1965). However, the position had some impact on scientific practice. For instance, the British astrophysicist Edward A. Milne (1896–1950), who was a pupil of Eddington, adhered to this position and as a result rejected curved space and the expanding universe. This sparked a major controversy in cosmology in the early 1930s with those who took curved space for a fact (cf. Gale and Urani 1993). In the mid-1930s, the allegiance to Milne's philosophical and scientific outlook furthermore led Howard Percy Robertson (1903–1961) and Arthur Geoffrey Walker (1909–2001) both independently to the development of a cosmological model—what is now known as the Robertson-Walker metric.

The other tradition that deals explicitly with the connection between theoretical and observational terms is that of Logical Empiricism, and in particular the work of Rudolf Carnap (1891–1970). In his *Foundations of Logic and Mathematics*, for instance, he distinguishes between knowledge in mathematics and knowledge in empirical science (Carnap 1939, p. 48) by highlighting that knowledge in mathematics does not refer to facts—that is, has no factual content (ibid., p. 55). Logic and mathematics nonetheless play an important role in physical science—namely, "to guide the deduction of factual conclusions from factual premisses" (ibid., p. 35). Calculus is widely used in physics, yet it can be formally separated from anything to do with factual content: there is (uninterpreted) calculus and there is the interpreted system (ibid., p. 57). The big challenge then is to give an account of how the calculus is interpreted. This is in effect the question of the connection between theoretical and observational terms. In Carnap's terminology of 1939, only some of the *signs* occurring in the calculus have *semantical rules* that connect them to things observable —which means they are *physical* or *elementary terms*: "we have to give semantical rules for elementary terms only, connecting them with observable properties of things" (ibid., p. 62).

However, other signs in the calculus lack this connection to observable properties of things. They can at best be indirectly—that is, *not explicitly*—defined. These abstract terms can be included into the testing procedure only by reference to less abstract terms (ibid., p. 63). Thus while some of the terms of a theory, and even its axioms, may not have an explicit interpretation, the theory needs to confront observation (ibid., pp. 66–67). The empirical test then comes through successful predictions, not through attempting direct interpretations of the axioms. The (partial) interpretation of the calculus is achieved through elementary (or observational) terms that are connected to observational facts by semantical rules. Carnap (ibid., p. 68) wrote: "It is true a theory must not be a 'mere calculus' but possess an interpretation, on the basis of which it can be applied to facts of nature. But it is sufficient, as we have seen, to make this interpretation explicit for elementary terms; the interpretation of the other terms is then indirectly determined by the formulas of the calculus, either definitions or laws, connecting them with the elementary terms." So much on Carnap's way of tackling the issue of theoretical terms. Whether the framework is operationalist or Logical Empiricist, if theories are construed in an axiomatic manner, linking them up to what is empirically observed presents a significant conceptual problem.

4.5 The Context of Justification and the Rational Reconstruction of Science

As we have seen, the status of theoretical terms was one of the central issues in connection with axiomatically conceived theories. There is a specific approach to science connected to the Logical Empiricist tradition, the rational reconstruction of science, that is responsible for the fact that theories, not models, were taken to play the dominant part in the study of science. Models simply do not feature there. Correspondingly, Carnap attributed only a very minor role to models. He said: "It is important to realize that the discovery of a model has no more than an aesthetic or didactic or at best a heuristic value, but it is not at all essential for a successful application of the physical theory" (ibid., p. 68). In Carnap's account, the meaning and interpretation of a theory is in no way dependent on the application of models. This general outlook is founded on a distinction that Hans Reichenbach (1891–1953) popularized, the distinction between the *context of discovery* and the *context of justification.* This distinction aims at the separation of logic and psychology, with logic as a means for analyzing scientific theories. For Reichenbach (1938, p. 6), it is not "a permissible objection to an epistemological construction that actual thinking does not conform to it." The interest is *not* in thinking and the context of discovery, but in "a relation of a theory to facts, independent of the man who found the theory" (ibid., p. 382).

It is not how Maxwell found his famous equations—namely, by employing a mechanical model—that counts. The psychology of scientists that may account for the discovery process is an empirical project. Logical Empiricist philosophy, in turn, is concerned with science in a nonempirical manner, thus not with its psychology (Carnap 1934, p. 6). Karl Popper (1902–1994), who moved close to the Vienna Circle, also highlighted the need to avoid confusion between psychological and epistemological problems. He wrote: "The initial stage, the act of conceiving or inventing a theory, seems to me neither to call for logical analysis nor to be susceptible of it" (Popper [1935] 1959, p. 31).

As with Carnap, this is an issue that would require examination by empirical psychology. The epistemological problems, however, are about justifying the knowledge contained in theories as being founded on empirical tests—that is, as based on observational terms (see section 4.4). Rational reconstruction of theories would then require presentation of theory in a form that makes explicit how theoretical postulates are connected to observational terms or data. The question is: Under what circumstances is a theory epistemologically justified? Scientific theories postulate universal principles, and yet these principles are to be based on individual empirical facts (Carnap 1936–1937). How can this link be established? Rational reconstruction was intended as a path to the clarification of that link with which to show how, in a rational, sound way, universal principles can be justified empirically. Thus it is knowledge that is to be put on a reliable base by reconstruction, not by the psychological process of discovery.

Once the project of Logical Empiricism is identified (Giere 1996), the almost nonexistent interest in modeling no longer comes as a surprise. As Reichenbach put it succinctly, the *actual* thinking process is, in this research context, not a legitimate subject of epistemology. The need or desire for mechanical models is not what matters. Interestingly, in opposition to the Logical Empiricist outlook, it is later precisely the "actual thinking"—imagining and creative cognitive processes—that became central to the study of scientific models and their role in scientific discovery (see section 4.7 and chapter 5). At the time, however, with the chosen project of rational reconstruction of science, there was no space nor need for the consideration of scientific models.

4.6 The British Context for the Discussion of Models

Clearly, Logical Empiricism was not to prepare the ground for the study of scientific models. In this section, I illustrate that for those British philosophers who began to concern themselves with models, it was largely the hypothetico-deductive account of theories that served as "fighting ground" (Hesse 1953, p. 198; Braithwaite [1953] 1968; and Hutten 1954, p. 297). It would seem that the influence of, for example, Carnap on British philosophy of science was, at least at that time, small (Hesse 2001, personal communication). Instead, Bertrand Russell (1872–1970), although no longer active in the philosophy of mathematics, and Ludwig Wittgenstein (1889–1951) were influences on British philosophy of science. Popper, in turn, was obviously no mainstream British philosopher, but was more likely to have exerted some influence. He had given lectures at Cambridge, London, and Oxford on an extensive visit in 1935 and 1936, and from 1946 onward he taught at the London School of Economics. Hesse (1953, p. 198) herself quoted Popper as one of the proponents of the hypothetico-deductive method.

Following Hesse (1952; see also Hesse 1954, chapter 7, for a discussion), it looks like operationalism was also perceived as a major opponent worthy of discussion. In her critique of the limits of operationalism, she sets out by suggesting "[i]t has for some time been generally accepted in physics that concepts occurring in physical theories must be operationally defined" (Hesse 1952, p. 281). The issue is again that of theoretical terms, and Hesse arrives at the conclusion that there exist concepts in physical theories that cannot be defined operationally, discussing the example of Dirac's quantum theory. She does not deny that there can be operational definitions in narrow contexts, but she takes it as an important lesson from operationalism ("implicit in several remarks made by such writers as Bridgman, Dingle, Dirac and Eddington" [ibid., p. 294]) that a distinction must be made between concepts based on direct measurement and those that we would now call theoretical terms. Hesse (ibid., p. 294) commented: "[I]t seems that in order to understand the significance of physical theories we are led to make a distinction between two sorts of concepts—those which can be measured or calculated and therefore operationally defined, and those which enter into theories as part of the mathematical apparatus for correlating observations, and which exhibit significant analogies with more familiar processes." Her own proposal is to interpret theoretical terms in terms of an analogy to some other, more familiar process.

While operationalism was not the only approach that led to the recognition of difficulty with the link between theoretical and empirical components in a theory, it seems that it may have served as a not-to-be-neglected background for philosophers in the British context. One thing operationalism does do, in any case, is bring into focus experiment and observation. Wanting to define at least some terms operationally requires attention to the practical side of science in one's philosophical considerations, and in the practice of science, models cannot so easily be neglected. In this sense, the tradition of operationalism may have exerted a positive influence in taking seriously scientific practice.

The issue of empirical and theoretical terms was of general concern (cf. Hesse 1954, p. 134) and also led Richard Braithwaite (1900–1990) (1954) to a detailed appreciation of scientific models. According to him, this was a necessary response both to Russell's views of theories and to the criticism by Frank Ramsey (1903–1930) of Russell's views.[3] Critical issues were, for instance, whether "electron," being unobservable, yet displaying observable behavior, is a theoretical or an empirical concept and how the empirical and the theoretical component in it are linked. Braithwaite opposed the view that an unobservable entity, such as an electron, can be logically constructed out of observable entities in a meaningful way.[4] He justifies this criticism by pointing out the lack of scope for theory development that results, writing: "To treat theoretical concepts as logical constructions out of observable entities would be to *ossify* the scientific theory in which they occur: . . . there would be no hope of extending the theory to explain more generalisations than it was originally designed to explain" (Braithwaite 1954, p. 36).

To summarize, in the 1920s and 1930s, several philosophical traditions were intent on dealing with the empirical foundation of theoretical terms, most prominently Logical Empiricism and operationalism. The difficulties of such an undertaking had been noted well and truly by 1950. The singular concern with the context of justification under the rule of Logical Empiricism had furthermore cut out issues of theory development and scientific discovery. Mere prejudice against models, as voiced by Duhem, may have also played its part in the suppression of any consideration of scientific models in philosophy.

4.7 The Emergence of Models in the Philosophical Discussion

The early proponents of the uses and benefits of scientific models in Britain[5]—Richard Braithwaite, Mary Hesse, and Ernest Hutten—shared some concerns, although they apparently worked entirely separately (Hesse 2001, personal communication). Hutten worked as a physicist at the Royal Holloway College, part of the University of London, where Hesse had completed her Ph.D. in mathematics before she became a lecturer in mathematics at the University of Leeds, but she certainly does not recall much encounter, let alone a discussion of models. Only Braithwaite was, at the time, an established philosopher, at the University of Cambridge. Hesse seems to have established personal contact with him before 1953. Using scientific theories as hypothetico-deductive systems as a platform, all three addressed some of the issues that this account failed to address. Discussing scientific models, they picked up on some of the same issues, although generally with quite different emphases.

4.7.1 A formal approach

Braithwaite ([1953] 1968 and 1954) was, of the three proponents, the most committed to a formal reconstruction of scientific theorizing. He considers scientific theorizing as a task of deduction, whereby a calculus formally represents the deductive system of the theory. The calculus itself is uninterpreted, which has the practical advantage that the calculus is clearly laid out and deductions from it are not confused by individual examples, but can be carried out merely in the form of symbolic manipulations (Braithwaite [1953] 1968, p. 23). However, it is undetermined how the calculus can be interpreted. Interpretation means that the symbols of the calculus are given meaning in the light of the empirical data, the occurrence of which the theory needs to explain. The underlying image Braithwaite employs is the following: Imagine the development from premises (of the calculus representing a deductive system) to inferred conclusions as a movement from top to bottom. At the bottom, one finds "directly testable lowest-level generalizations of the theory" (Braithwaite 1954, p. 38).

In other words, observational data stand at the bottom of the logical chain of reasoning. The actual construction of theories, however, takes place precisely in the opposite direction—namely, from bottom to top. It starts with observational data at the bottom from which the premises or hypotheses at the top of the logical chain need to be found. Although the premises are "logically prior," they are "epistemologically posterior" (Braithwaite [1953] 1968, p. 89)—that is, in the actual development of the theory, observed events are known before any higher-level hypotheses can be known. The theory formulation confronts an epistemological problem, because the logically posterior consequences (the observational data) determine the meaning of the theoretical terms—that is, of the logically prior hypotheses in the calculus representation of the theory (ibid., p. 90).

To be able to work from the logically prior to the logically posterior —that is, from "top" to "bottom" —provisional or hypothetical interpretations of the calculus and of the premises in particular are required. These, according to Braithwaite, can be provided in full by models, because models have a different epistemological structure from theories. A model, in contrast to a theory, is an interpreted calculus.[6] In the model, the interpretation of the premises is fixed, even if hypothetically, while the model can still have the same structure as the theory. To illustrate the epistemological difference between model and theory, Braithwaite (ibid., p. 90) uses the metaphor of a zip: "[T]he calculus is attached to the theory at the bottom, and the zip-fastener moves upwards; the calculus is attached to the model at the top, and the zip-fastener moves downwards."

Because a model is fully interpreted, whereas a theory is not, the model is a more accessible way to think about the structure represented by the calculus, which makes the model an alternative way of thinking about the theory. For Braithwaite, epistemological advantages establish the role of models: the need to provide an interpretation of a calculus, at least hypothetically. The framework of Braithwaite's argument—and his conception of a model—is largely formal, very much in the Logical Empiricist tradition, but epistemology becomes a topic within this formal context for which it begins to matter whether "actual thinking conforms to it." He wrote: "What we are concerned with are straight logical problems of the internal structure of scientific systems and of the rules played in such systems by the formal truths of logic and mathematics, and also the problems of inductive logic or epistemology concerned with the grounds for reasonableness or otherwise of accepting well-established scientific systems" (ibid., p. 21).

Comparing Braithwaite's with Carnap's position in the 1930s, for both an uninterpreted calculus features that requires interpretation by means of empirical data that is connected to the conclusions inferred from the calculus, but Carnap simply states that the premises, the axioms, of the theory are not—certainly not directly—interpreted. Braithwaite, however, chooses a different route and introduces hypothetical interpretations of the calculus: namely, models. Braithwaite also considers the practical concerns of developing a theory. For Braithwaite, it matters that an only partially interpreted theory is hard to think in terms of, whereas accessibility can be achieved with a hypothetical interpretation. Braithwaite (ibid., p. 368) shows considerable awareness for the issues of "actual thinking," even if he studies them in a formalistic manner: "The business of a philosopher of science is primarily to make clear what is happening in scientific thinking." Here, the distinction between context of discovery and context of justification (section 4.5) is distinctly softened.

4.7.2 A practical approach

Hesse and Hutten departed from Braithwaite's formalism. They focus yet more on models, rather than theory, and developed their conceptions of scientific models more closely from scientific practice and from the actual needs of scientists, also actual needs for thinking—while they continue to work on the assumption that theories are hypothetico-deductive. For Hutten (1954, p. 285), it is no good if accounts of scientific method differ hugely between scientists and philosophers, and "model" is just such a term that features in scientists' understanding of scientific method yet is totally neglected by philosophers. Hutten (1956, p. 81) advises: "It is obviously best to follow the scientists here as closely as possible, at least in the first instance; we may hope in this way to avoid forcing science into a pre-conceived scheme, as philosophers have so often done." Part of this endeavor is to "take as an example a modern theory and discuss actual laws, instead of illustrating scientific method by means of old-fashioned and very simplified examples" (Hutten 1954, p. 284).

Hutten himself discusses, among other things, the model of an oscillator applied to the specific heat of solids and other areas of physics. With this, he anticipates the importance of case studies and examples for an exploration of scientific models. Hesse's (1953) article does in fact contain a case study to illustrate her claims about models. She discusses the development of various nineteenth-century models of the transmission of light in the aether. Employing such a case study is used as a new type of argument that enters the philosophical discussion and carries considerable weight in demonstrating the need for the study of models in philosophy of science.

Hutten describes scientific models as "partial interpretations" of theories in the sense that models do not aim to be a *copy* of the theory nor of reality.[7] He says: "[T]here is always some element that is changed or left out in the model as compared to the thing of which it is a model" (Hutten 1954, p. 286). This means that models can be misleading; moreover, their status is that of being neither true nor false (ibid., p. 296), and there can even exist multiple models of the same thing. He continues: "[W]e may have many auxiliary models within a single theory; usually, they overlap and are mutually compatible though, on occasion, the models are alternatives" (ibid., p. 298).

In Hesse's article, Hutten's notion of partial interpretation is paralleled by the claim that models cannot be regarded as "*literal descriptions of nature, but as standing in a relation of analogy to nature*" (Hesse 1953, p. 201; italics in original). The point here is that the model only describes certain aspects of a phenomenon in nature, but not others. The model may even misdescribe certain aspects (for example, disanalogies to nature). This is why Hutten says that models can mislead and why he calls them neither true nor false. He makes numerous perceptive points about scientific models. Although Hutten does not elaborate on them in great detail nor with philosophical rigor, they are precursors of what I examine in chapters 5, 6, and 8 as central areas of investigation concerning models. This makes Hutten a valuable source of those issues that must be counted among the motivations for the study of models. For instance, he is one of the first as a philosopher-physicist to talk explicitly and positively about a psychological function of models.

This "heuristic" or "pragmatic" use of models is based on the fact that models provide a *visual representation* of something. They do so either in three dimensions or in two dimensions in the form of pictures or diagrams (Hutten 1954, p. 285). Models, Hutten proposes, make available appropriate vocabulary, another practical issue in constructing scientific theories: "The model prescribes a context, or gives a universe of discourse" and it "supplies primarily a terminology" (ibid., p. 295). Being confronted with a new situation, orientation comes from comparing this new situation with a familiar situation. He writes: "In science, we merely want to explain a new and unfamiliar phenomenon, and so we try to account for it in terms of what we already know, or to describe it in a language that is familiar to us. That is, the model is used to provide an *interpretation* [of the new phenomenon]" (ibid., p. 286).

Incidentally, familiarity is precisely not what is required for an explanation, according to Hempel and Oppenheim ([1948] 1965, p. 257). For Hutten, in turn, both familiarity with phenomena and with language for the purposes of description matter. In this vein, he also compares models with metaphor (cf. chapter 5, section 5.4). Models function like metaphors because they are used "when, for one reason or another, we cannot give a direct and complete description in the language we normally use" (Hutten 1954, p. 298; see also ibid., p. 293) —that is, when easily available terminology fails us. There are some hints that models are supposed to serve as a link between theory and experiment (cf. chapter 6). According to Hutten (ibid., p. 289), theories are explained and tested in terms of models, although he does not specify how. Somewhat obscurely, he states that the model is not an application of the theory, although the theory is applied with its help.[8]

An important aim of Hesse's article is to expand the concept of model to go beyond purely mechanical nineteenth-century models. She argues that mathematical formalisms can also be scientific models and that no sharp line should be drawn between the two because they function in essentially the same way (Hesse 1953, p. 200). For example, the mathematical development of a model may suggest empirical results that have not previously been observed, such as Fresnel's spot, as described in section 4.1. Allowing for "mathematical models" is an important step toward a wider use of the concept of model, later taking effect in the notion of a theoretical model (for example, Achinstein 1965).

Like Braithwaite, Hesse considers theories as hypothetico-deductive. Although this indicates a framework of discussion similar to Braithwaite's, Hesse highlights an entirely different point arising from a situation where inferences are required from "bottom" (empirical data) to "top" (hypotheses of the theory). She writes: "The main point that emerges from such a description of theories is that there can be no set of rules given for the procedure of scientific discovery—a hypothesis is not produced by a deductive machine by feeding experimental observations into it: it is *a product of creative imagination, of a mind* which absorbs the experimental data *until it sees them fall into a pattern,* giving the scientific theorist the sense that he is penetrating beneath the flux of phenomena to the real structure of nature" (Hesse 1953, p. 198; italics mine). This is where models come in, as tools for the creative process of discovery and theory development. Hesse here anticipates her own future philosophical concerns, exploring procedures for scientific discovery and creative imagination, in which models become central players. Moreover, she makes explicit reference to *actual thinking,* the mental activity of scientists.

4.8 The Modelers as Visionaries

The work of Hutten, Hesse, and Braithwaite in the early to mid-1950s first properly recognized scientific models as a topic worthy of study. As we have seen, the following issues inspired their interest:

- models for the purposes of interpreting theory (cf. chapter 6, section 6.5);
- how scientists use models (cf. chapter 6, section 6.2);
- models as incomplete and "not literal"—that is, as leaving out things and potentially being misleading (cf. chapter 8);
- models as aids for visualization;
- models as relating to the familiar (cf. chapter 3);
- models as providing descriptive vocabulary (cf. chapter 5, section 5.3);
- models as guides for experimentation;
- models as tools for thinking and theory development (cf. chapter 5, especially section 5.4.2).

Models were also seen as one way of bridging the gap between theory and observation, between theoretical and observational terms. The question of linking theory with observation language was soon to be revolutionized by Norwood Russell Hanson (1924–1967) (1958), and the discussion of theory change was further sparked by Thomas Kuhn's (1962) concept of scientific revolution. The related issues of scientific discovery and creative imagination continue to be important concerns. Similarly, the argument that models rely on the familiar to make accessible the unfamiliar remains prominent (see chapters 3 and 5). The proposal that scientific models should be studied considering scientific practice has also, since these early days of model consideration, fallen onto fertile ground (see chapter 6). The notion that models are not copies and not literal descriptions of nature is found again in the recent debate about representation and realism of scientific models (see chapter 8). Finally, Hesse's notion that models need not be mechanical, or that what counts as mechanical must not be determined too narrowly, also has unexpected repercussions in current discussions on mechanisms in science (see chapter 2, section 2.4). Once awareness of such central themes, originally touched upon in a small number of articles in the 1950s, was raised, models entered a phase of great popularity. The 1960s saw an enormous proliferation of articles in which a growing number of philosophers tried to establish what exactly the role of models was in science. Theories had not disappeared from the stage, but models had established themselves beside them.

Notes

1. This is a problem to which models were later seen as a solution, models as interpretations of abstract theories.

2. Duhem's concept of explanation clearly differs from what later gained currency as the deductive-nomological account of explanation. For Duhem ([1914] 1954, p. 7), the difficulty with explanation is that giving an explanation would require secure knowledge of "what things are like" under the veil of appearances: "To explain (explicate, *explicare*) is to strip reality of the appearances covering it like a veil, in order to see the bare reality itself." In contrast to this, the status of theories is such ("descriptive"), according to Duhem, that they can no more than provide hypothetical explanations—that is, do not make definitive claims about the structure of reality. It therefore strikes me that Duhem, when he calls models "explanatory" in the previous quote, does not mean explanatory in the sense of "providing an explanation" in the narrow, metaphysical sense spelled out in his chapter 1, but perhaps in the sense of "creating a visible and palpable image," as befits his account of models in chapter 4.

3. For a discussion of Russell's theory of logical construction and Ramsey's refutation, see Hesse 1954, pp. 109–113.

4. For a detailed discussion of this point, see Braithwaite 1953, chapter 3.

5. There are a number of publications on models in the American journal *Philosophy of Science* in the late 1940s and early 1950s. They seem not to have made any impact of note, however, in that they are rarely, if ever, cited. With the exception of Rosenblueth and Wiener (1945), they appear to originate in a German tradition, quoting Kant (Altschul and Biser 1948) or Goethe's *Faust* in German besides Kant, Hegel, and Marx (Deutsch 1951), presumably by German immigrants to the United States.

6. The model is not that by which the calculus is interpreted, as the Semantic View of theories would have it (see chapter 6, section 6.1), but the calculus interpreted—that is, the theory including an interpretation of the theory.

7. This is not the same use as Carnap's use of the term "partial interpretation."

8. Hesse (1953, p. 199), in turn, endorses the link between models and experiment with a slightly different emphasis: "Progress is made by devising experiments to answer questions suggested by the model."

References

Achinstein, P. 1965. Theoretical models. *British Journal for the Philosophy of Science* 16: 102–120.

Altschul, E., and E. Biser. 1948. The validity of unique mathematical models in science. *Philosophy of Science* 15: 11–24.

Braithwaite, R. [1953] 1968. *Scientific Explanation: A Study of the Function of Theory, Probability, and Law in Science.* Cambridge: Cambridge University Press.

———. 1954. The nature of theoretical concepts and the role of models in an advanced science. *Revue Internationale de Philosophie* 8 (fasc. 1–2): 34–40.

Bridgman, P. [1927] 1961. *The Logic of Modern Physics.* New York: Macmillan.

Carnap, R. 1934. On the character of philosophic problems. *Philosophy of Science* 1: 5–19.

———. 1936–1937. Testability and meaning. *Philosophy of Science* 3: 419–471 and 4: 1–40.

———. 1939. Foundations of logic and mathematics. In *International Encyclopedia of Unified Science.* Chicago: University of Chicago Press, pp. 1–70.

Danto, A., and S. Morgenbesser. 1960. *Philosophy of Science.* New York: New American Library.

Deutsch, K. W. 1951. Mechanism, organism, and society: Some models in natural and social science. *Philosophy of Science* 18: 230–252.

Duhem, P. [1914] 1954. *The Aim and Structure of Physical Theory.* Translated from the French 2nd edition. Princeton, New Jersey: Princeton University Press.

Eddington, A. [1923] 1975. *The Mathematical Theory of Relativity.* New York: Chelsea Publishing Company.

Fölsing, A. 1997. *Heinrich Hertz: Eine Biographie.* Hamburg: Hoffmann und Campe.

Gale, G., and J. Urani. 1993. Philosophical midwifery and the birthpangs of modern cosmology. *American Journal of Physics* 61: 66–73.

Giere, R. 1996. From wissenschaftliche Philosophie to philosophy of science. In R. Giere and A. Richardson, eds., *Origins of Logical Empiricism.* Minnesota Studies in the Philosophy of Science, vol. 16. Minneapolis: University of Minnesota Press, pp. 335–354.

Gower, B. 1997. *Scientific Method: An Historical and Philosophical Introduction.* London: Routledge.

Hanson, N. R. 1958. *Patterns of Discovery.* Cambridge: Cambridge University Press.

Harman, P. M. 1982. *Energy, Force, and Matter: The Conceptual Development of Nineteenth-Century Physics.* Cambridge: Cambridge University Press.

Hempel, C. G. [1954] 1965. A logical appraisal of operationalism. In *Aspects of Scientific Explanation.* New York: The Free Press, pp. 123–133.

———, and P. Oppenheim. [1948] 1965. Studies in the logic of explanation. In C. G. Hempel, *Aspects of Scientific Explanation and Other Essays in the Philosophy of Science.* New York: The Free Press, pp. 245–290.

Hertz, H. [1890] 1984. Über die Grundgleichungen der Elektrodynamik für ruhende Körper (On the fundamental equations of electrodynamics for stationary bodies). In *Gesammelte Werke, Vol. 2: Untersuchungen über die Ausbreitung der elektrischen Kraft* (Collected works, vol. 2: Investigations of the propagation of the electrical force). Vaduz: Sändig-Reprint-Verlag, pp. 208–255.

Hesse, M. 1952. Operational definition and analogy in physical theories. *British Journal for the Philosophy of Science* 2: 281–294.

———. 1953. Models in physics. *British Journal for the Philosophy of Science* 4: 198–214.

———. 1970. Is there an independent observation language? In R. G. Colodny, ed., *The Nature and Function of Scientific Theories.* Pittsburgh: University of Pittsburgh Press, pp. 35–77.

Hesse, M. B. 1954. *Science and the Human Imagination.* London: SCM Press.

Hutten, E. 1956. *The Language of Modern Physics.* New York: Macmillan.

Hutten, E. H. 1954. The role of models in physics. *British Journal for the Philosophy of Science* 4: 284–301.

Kuhn, T. S. 1962. *The Structure of Scientific Revolutions.* Chicago: University of Chicago Press.

O'Hara, J. G., and W. Pricha. 1987. *Hertz and the Maxwellians.* London: Peregrinus.

Popper, K. [1935] 1959. *The Logic of Scientific Discovery.* London: Hutchinson.

Reichenbach, H. 1938. *Experience and Prediction.* Chicago: University of Chicago Press.

Rosenblueth, A., and N. Weiner. 1945. The role of models in science. *Philosophy of Science* 12: 316–321.

5 PARADIGMS AND METAPHORS

MAKING SCIENTIFIC PROGRESS often requires thinking about a phenomenon in a novel manner. There exist at least a couple of articles, published in *Philosophy of Science* in 1951, that emphasize the role of models for thinking. Herman Meyer (1951) more or less equates models with *mental pictures* and views these mental pictures as a route toward linking mathematical expressions to observations. He writes: "Many people ask for something that makes the meaning of those formulae *intuitively clear*. They want a 'mental picture' re-establishing the connection between the perfectly abstract and often very abstruse mathematical formulae and direct observations made in laboratories and in nature generally" (ibid., p. 112). The understanding is that models make ideas embodied in theories intuitively clear (ibid., p. 113). Meyer (ibid., p. 118) proposes that more can be achieved than mathematical descriptions of observations: "Our specifically human way of progressing from phenomenological description to *scientific knowledge*, is a mental operation *sui generis*, which I have called here the construction of scientific models."

In a similar spirit, Karl Deutsch (1951) begins his article "Mechanism, Organism, and Society" with the following strong statement: "Men think in terms of models. Their sense organs abstract the events which touch them; their memories store traces of these events as coded symbols; and they may recall them according to patterns which they learned earlier, or recombine them in patterns that are new" (ibid., p. 230). To me, these quotes from Meyer and from Deutsch with the intuitions expressed sound somewhat reminiscent of the ideas behind notions, such as paradigms or metaphors, or even the mental models literature beginning in the 1980s (for example, Gentner and Stevens 1983; and Johnson-Laird 1983). To my knowledge, these early authors on models did not, however, have any impact on the discussion of modeling commencing in the 1950s discussed in chapter 4 (section 4.7).

To make scientific progress, sometimes a transition needs to be made from one way of thinking about something to another. Sometimes, two ways of thinking may be exemplified by different models, such as the Ptolemaic and the Copernican models of the solar system. This kind of transition has been referred to as "paradigm change" by Thomas Kuhn (1922–1996) in his 1962 *The Structure of Scientific Revolutions* (section 5.1). Another approach to models that picks up this idea of conceptual change, although in a different manner, is comparing models to metaphors. Well before Kuhn's seminal work, metaphor started to be used in philosophy of science (Hesse 1953; Hutten 1954; and Harré 1960), albeit not in a systematic manner. Thinking about metaphor then became a way of thinking about scientific discovery processes and conceptual change. Metaphors were postulated to have "cognitive content" (Black 1962; and Hesse 1963b). To exploit the analogy between metaphor and model, it is indispensable to consider the interaction view of metaphor, as done in section 5.2. One issue is what we can learn from the analogy between models and metaphors, another is the separate issue of how metaphorical language is used in science and scientific modeling. The latter is the topic of section 5.3. I consider in section 5.4 what exactly can be learned from metaphor when studying scientific modeling, and what role metaphor plays in modeling. Section 5.5 contains the conclusion.

5.1 Paradigms

In the preface of *The Structure of Scientific Revolutions*, Kuhn ([1962] 1970, p. viii) introduces paradigms as "universally recognized scientific achievements that for a time provide model problems and solutions to a community of practitioners." Often these universally recognized scientific achievements can be found represented in terms of science textbooks. Textbooks introduce students to typical examples and typical solutions to example problems. (Kuhn later calls these "exemplars.") Learning example problems and solutions is required in order to join professionally the practitioners of a scientific discipline: "The study of paradigms . . . is what mainly prepares the student for membership in the particular scientific community with which he will later practice" (ibid., p. 11).

According to Kuhn, learning to solve problems is to see how a problem solution is similar to the solution of a problem previously solved. Kuhn says, in *Second Thoughts on Paradigms*, that "a perception of similarity . . . is both logically and psychologically prior to any of the numerous criteria by which that same identification of similarity might have been made. . . . The mental or visual set acquired while learning to see two problems as similar can be applied directly" (Kuhn [1974] 1977, p. 308). It is this "mental or visual set" that comes closest to the notion of "models in terms of which we think," in contrast to some other uses of the term "paradigm" that are not cognitive and more sociological. Giving the example of little Johnny learning to distinguish ducks, geese, and swans, Kuhn (ibid., p. 313) emphasizes that "shared examples have essential cognitive functions prior to a specification of criteria with respect to which they are exemplary."

One may object, of course, that the analogy between little Johnny learning names of birds and a student learning advanced physics may have its drawbacks, just as more may be involved in a paradigm change than the gestalt switch from seeing a drawing of a duck to seeing it as a rabbit (cf. Nickles 2003, p. 164). It is surprising, given his cognitive views, that Kuhn never systematically linked up his claims with findings from cognitive psychology, such as schema theory or work on mental models, or with computational work on rule-based and case-based reasoning. After all, many of Kuhn's claims appear largely to be in agreement with findings in these areas (Nersessian 2003; and Nickles 2003).

Kuhn's use of the concept of a paradigm is riddled with difficulties in that there are a number of different uses of the term (Shapere 1964; and Masterman 1970). Kuhn himself acknowledges two different uses of "paradigm" in his *Postscript* to *Structure*, "the entire constellation of beliefs, values, techniques, and so on shared by the members of a given community," then called "disciplinary matrix," and "the concrete puzzle solutions," called "exemplars" (Kuhn 1970, p. 175; see also Kuhn [1974] 1977). An important use in the current context is a paradigm as a way of thinking about a set of problems that does not, at the time, require further justification because it is accepted by the scientist's fellow practitioners. Usually, the paradigm is accepted because it does solve certain problems and is in that sense potentially better than its competitors, but this does not mean that the paradigm is capable of solving all problems arising in a field (Kuhn [1962] 1970, pp.17–18 and 23). The paradigm can also specify the problem that needs to be solved and in that sense influence the kind of experiments conducted or the apparatus used (ibid., p. 27).

The idea of a paradigm is that it can guide research—that is, raise problems—despite not being reducible to rules (ibid., p. 44). A paradigm is somehow intuitive rather than rule-driven. Depending on their experience, scientists can "see" different things. A student may see lines on paper where a cartographer sees a picture of a certain terrain (ibid., p. 111). It would appear here that Kuhn mixes a way of thinking about a subject with the perceptual experience of it. We may *see* things in different ways, just as we may think about and interpret them differently, depending on our paradigm. After all, Kuhn's image for the paradigm shift is Ludwig Wittgenstein's example of the duck-rabbit, where lines on paper can be seen as either a duck or a rabbit but not both (Wittgenstein 1953, p. 194). A change of paradigm is compared to a gestalt shift, and Kuhn ([1962] 1970, p. 116) also talks about "paradigm-induced changes in scientific perception"(see also Hanson 1958, pp. 11ff.).[1]

Models are not the same as paradigms, but they contribute to the paradigm. In his *Second Thoughts on Paradigms*, Kuhn mentions models as one constituent, besides, for example, symbolic generalizations and exemplars, of a disciplinary matrix. He says about models: "Models . . . are what provide a group with preferred analogies or, when deeply held, with an ontology. On one extreme they are heuristic: the electric circuit may fruitfully be regarded as a steady-state hydrodynamic system, or a gas behaves like a collection of microscopic billiard balls in random motion. On the other, they are the objects of metaphysical commitment: the heat of a body *is* the kinetic energy of its constituent particles, or, more obviously metaphysical, all perceptible phenomena are due to the motion and interaction of qualitatively neutral atoms in the void" (Kuhn [1974] 1977, pp. 297–298). The mention of analogy perhaps indicates that the transition from Kuhnian paradigms to scientific models as metaphors is not as stark as it might first appear. Even if not in terminology, there are strong affinities in spirit between paradigm-driven and metaphor-driven views.[2]

Moreover, in his "Metaphor in Science," Kuhn ([1979] 1993, p. 538) comments: "Though not prepared here and now to argue the point, I would hazard the guess that the same interactive, similarity-creating process which [Max] Black has isolated in the functioning of metaphor is vital also to the function of models in science." The view of metaphor by Max Black and others is the subject of the next section. The thread between these different sections remains the notion of some cognitive element underlying scientific modeling.

5.2 The Metaphor Approach to Models

The idea of viewing scientific models as metaphors appeared in the 1950s (Hesse 1953; Hutten 1954; see also Bailer-Jones 2002) and was taken up in Mary Hesse's (1963a) work in the 1960s with the aim to show that metaphorical models and analogy are more than heuristic devices that can be jettisoned once a "proper" theory is in place.[3] Work on models and metaphor continues to be discussed to this day (Paton 1992; Bhushan and Rosenfeld 1995; Miller 1996; Bradie 1998 and 1999; and Bailer-Jones 2000), and it usually concurs with the rejection of the dominance of theories over models—a rejection that need not be linked to a metaphor view of models, however (see chapter 7). Metaphor, in this context, is seen as closely tied to analogy, just as models are closely tied to analogy (Achinstein 1968; Harré 1988; cf. chapter 3). The claim that scientific models are metaphors is tied to the fact that often an analogy is exploited to construct a model of a phenomenon. Thus, if scientific models are metaphors, then analogy is an important factor in this. "The brain is the hardware for which a child gradually develops suitable software" implies an analogy between data processing in a computer and the cognitive development of a child, just like the liquid drop model of the atomic nucleus suggests an analogy between the atomic nucleus and a liquid drop in that the overall binding energy of the nucleus is, in approximation, proportional to the mass of the nucleus—like in a liquid drop. The view that scientific models are metaphors depends, of course, on what metaphor is taken to be, besides metaphors being suggestive of analogies.

Although the idea that scientific models are metaphors appears in Black 1962, it was further explored in Hesse (1963) 1966.[4] A metaphor is a linguistic expression in which at least one part of the expression is transferred from one domain of application (source domain), where it is common, to another (target domain) in which it is unusual, or was probably unusual at an earlier time when it might have been new.[5] This transfer serves the purpose of creating a specifically suitable description of aspects of the target domain, where there was no description before (for example, "black hole") or none was judged suitable. It is often presumed that the metaphorical phrase has a special quality in the way it communicates information, sometimes referred to as "cognitive content" (Black 1954).[6] A "strong cognitive function" is assigned to metaphor when "a metaphorical statement can generate new knowledge and insight by *changing* relationships between the things designated (the principal and subsidiary subjects)" (Black [1979] 1993, p. 35).

This is thought to happen because metaphor inspires some kind of creative response in its users that cannot be rivaled by literal language use. Think of "little green men" as a metaphor for extraterrestrial intelligent life, as it is used in science, not restricted to fantasy. Of course, the original domain of that expression is fantasy, and there "little green men" may mean exactly that: small green people. If the phrase is used in science contexts, however, the implicit reference to fantasy highlights the fact that we have no idea of what extraterrestrial intelligent life might be like. Something naively and randomly specific—little green men—is chosen to indicate that there is no scientific way of being specific about the nature of extraterrestrial intelligent life. Precisely the fact that we do not know what extraterrestrials are like is what we grasp from the phrase "little green men."

No amount of interpreting the literal phrase without a system of "associated commonplaces" (Black 1954; "implicative complex" in Black [1979] 1993) would enable us to achieve this; thus knowledge of the domain of application is crucial. "Little green men" is transferred from the domain of fantasy to the radically different domain of science, where one would not usually expect this expression. Nonetheless, there is no reason to think that the metaphor of little green men can be less reliably interpreted and understood by its recipients than any phrase from the "right" domain of language use, such as "not further specifiable forms of extraterrestrial intelligent life."

On the contrary, according to Max Black's interaction view, which goes back to the work of Ivor Richards (1936), we even gain insight through the metaphor that no literal paraphrase could ever capture; a metaphor cannot be *substituted* by a literal expression. Neither is it simply a *comparison* between the two relevant domains, as in an elliptical simile ("Extraterrestrial intelligent life is [like] little green men"), because, as Black suspects, metaphor can *create similarity.*[7] If this is true, metaphorical meaning can no longer be viewed as a sheer function of the literal meaning of linguistic expressions belonging to a different domain. Instead, the proposal of the interaction view is that the meanings of the linguistic expressions associated with either domain shift. The meanings of the expressions are extended due to new ideas that are generated when the meanings associated with source and target subject interact. The interaction takes place on account of the metaphor, which forces the audience to consider the old and the new meaning together.

Hesse (1966, pp. 158–159) draws from the interaction view of metaphor in her discussion of scientific models: "In a scientific theory the primary system is the domain of the explanandum [target domain], describable in observation language; the secondary is the system, described either in observation language or the language of a familiar theory, from which the model is taken [source domain]: for example, 'Sound (primary system) is propagated by wave motion (taken from a secondary system),' 'Gases are collections of randomly moving particles.'" Hesse goes on to postulate a meaning shift for metaphors. She takes the shift to take place with regard to pragmatic meaning that includes reference, use, and a relevant set of associated ideas (ibid., p. 160). Correspondingly, a shift in meaning can involve a change in associated ideas, a change in reference, and/or a change in use. On these grounds, Hesse gets close to dissolving the literal/metaphorical distinction. She writes: "[T]he two systems are seen as more like each other; they seem to interact and adapt to one another, even to the point of invalidating their original literal descriptions if these are understood in the new, postmetaphoric sense" (ibid., p. 162; cf. Hesse 1983).

The crucial point is that metaphors can (in spite of or because of this) be used to communicate reliably and are not purely subjective and psychological. Not "*any* scientific model can be imposed a priori on *any* explanandum and function fruitfully in its explanation" (Hesse 1966, p. 161). Scientific models, in contrast to poetic metaphors, are subject to certain objective criteria, or as Hesse (ibid., p. 169) puts it: "their truth criteria, although not rigorously formalizable, are at least much clearer than in the case of poetic metaphor." Correspondingly, one may "[speak] in the case of scientific models of the (perhaps unattainable) aim to find a 'perfect metaphor,' whose referent is the domain of the explanandum" (ibid., p.170).

In my formulation, a model is evaluated regarding whether it provides access to a phenomenon and matches the available empirical data about the phenomenon reasonably well (cf. chapters 1 and 8). The conceptual difficulty one confronts when considering whether scientific models are metaphors is that what makes a metaphor insightful is the analogy that it suggests. As we have seen in chapter 3, Hesse spends a number of chapters of her book *Models and Analogies in Science* discussing analogy. So one might argue that insights that are claimed to be due to metaphor may just as well be said to be due to analogy. Thus what exactly is the relationship between metaphor and analogy with respect to models? The relationship of analogy is usually an important factor in being able to understand a metaphor. Yet establishing the importance of analogy for understanding a metaphor is not to claim that a metaphor is only formulated once an analogy has been recognized. One could equally argue that it is the metaphor that prompts the recognition of an analogy—and it is feasible that *both* types of cases occur; the latter possibility would still warrant that the metaphor is connected to the analogy (or analogies) suggested by it ("every metaphor may be said to mediate an analogy or structure correspondence," in Black [1979] 1993, p. 30).

In astronomical observations, one talks about the signal-to-noise ratio. *Signal* is the light emitted from the object one wants to observe; *noise* represents the uncertainty in the signal (and the *background*) due to quantum fluctuations of photon emission and thus represents a limit to the precision with which the signal can be determined. The original metaphor connected with the noise metaphor is to a sound signal—for example, emitted from an interlocutor—while noise from other people talking and perhaps a nearby road needs to be separated from the signal in order to make out the information of interest. As listeners dealing with sound waves, we are quite proficient in filtering out all those unpredictable random sources that could prevent us from detecting the signal in which we are interested, and a comparable skill would be required for optical waves in astronomy. Without this analogy, the metaphor of noise, as used in astronomy, is incomprehensible.

5.3 Metaphorical and Literal Scientific Vocabulary

The analysis of metaphor is traditionally conducted by contrasting literal with figurative (or metaphorical) language. This requires the reliance on an intuitive or commonsense understanding of "literal," despite the difficulty of pinpointing what makes literal language literal. Of course, we have a sense in which talk about "little green men" appears metaphorical in comparison to "extraterrestrial intelligent life." "Literal" implies, by default, that an expression is not transferred from another domain—that is, "more directly" about something and perhaps more "typical," "common," "usual," or "expected." Inevitably, such a classification remains unsatisfactory, partly because we do not tend to find metaphorical statements more difficult to comprehend than so-called literal statements that could stand in their place (Rumelhart [1979] 1993). Metaphors, moreover, can be perfectly usual and familiar. Nobody stumbles over *processing information* or *developing software* said of the mind, or a *phylogenetic tree* merely because it is no oak, beech, lime, or fir. Just as we understand the brain-as-computer metaphor, we understand that a phylogenetic tree displays dependency relations of a group of organisms derived from a common ancestral form, with the ancestor being the trunk and organisms that descend from it being the branches. Most metaphors are understood with ease, which indicates that there are no grounds to treat them as deviations of language use. On the contrary, they are pervasive and central (Richards 1936).

Although there may be no clear-cut distinction between literal and metaphorical, one can still observe different degrees of metaphoricity, and the conditions under which we are capable of comprehending metaphors can be outlined correspondingly. The following are examples of such different degrees of metaphoricity:

- Even though a metaphor is entirely novel to us, we are endowed with the cognitive skill to interpret it just as easily as if we were familiar with that particular use of terminology.
- While we recognize a phrase as metaphorical in principle, we are so familiar with the particular type of metaphor that the metaphor is neither unusual nor unexpected; the brain-as-a-computer metaphor is an example of this. Another is to think of the energy distribution of a system as a landscape with mountains and valleys, and a gravitational force that is responsible for differences in potential energy depending on height, exemplified in such phrases as *potential well* or *tunnelling through a potential barrier*.
- We are so familiar with what once was a metaphor that a special effort would be required to recognize it as such; examples are electric *current*, electric *field*, *excited* state, or a chemical *bond forming*, *breaking*, *bending*, *twisting*, or even *vibrating*. Such metaphors are "dead"; they are pervasive in our language and they appear to us just like literal expressions (Machamer 2000), especially as sometimes they are our *only* expression for what they describe. Historical priority would probably be the only grounds on which a *current* of a river or a *field* plowed by a farmer would be judged more literal than *electric current* or *electric field*.

These degrees of metaphoricity are only partially related to the novelty of the metaphors, because some metaphors (unlike those in the third or even in the second example above) will always remain recognizable as metaphorical, no matter how familiar and well known they have become. An example would be "God does not play dice" expressing resistance to indeterminacy in physics. Rom Harré and his coauthors discuss this relationship between models and metaphor. They claim that both could be interpreted successfully with the same tool—namely, their type-hierarchy approach (Aronson, Harré, and Way 1995, pp. 97ff.). Yet the role of metaphor in science is separate again (Harré 1960 and 1970, p. 47; Martin and Harré 1982).

According to J. Martin and Rom Harré (1982), metaphorical language is used in the sciences to fill gaps in the scientific ordinary language vocabulary. Examples are metaphorical expressions that have acquired specific interpretations, like *electric field*, *electric current*, or *black hole*, which is why they should be understood "without the intention of a point-by-point comparison" (ibid.). Such metaphorical terms are a spin-off of scientific models (ibid., p. 100). Martin and Harré (ibid., p. 100) explain: "The relationship of model and metaphor is this: if we use the image of a fluid to explicate the supposed action of the electrical energy, we say that the fluid is functioning as a model for our conception of the nature of electricity. If, however, we then go on to speak of the 'rate of flow' of an 'electrical current,' we are using metaphorical language based on the fluid model."

It seems that many examples lend force to the view that metaphorical scientific terminology, even if hardly recognizable as such any longer, can be a spin-off of models (without the claim that models themselves are metaphorical). That Martin and Harré, different from myself, consider models simply as analogues has no bearing on this specific point. In many instances of metaphorical terminology in science, this vocabulary is employed to meet the problem of catachresis—that is, to provide scientific terminology where none existed previously (Boyd 1993). In chapter 3, section 3.4, I introduced the analogy of simulated annealing, an optimization technique for determining the best fit parameters of a model that is based on some data. In the computational method of simulated annealing, not only the equations from statistical physics (such as the Boltzmann equation) are adopted almost exactly, but the descriptive terminology is also taken over. Such terms as *temperature*, *specific heat capacity*, and *entropy* are applied to optimization in a meaningful way (Bailer-Jones and Bailer-Jones 2002).

The use of metaphorical vocabulary in science is not the same issue as the claim that models are metaphors. I discussed metaphorical vocabulary for two reasons: first to distinguish the two issues clearly, and second to highlight that metaphorical vocabulary can be a spin-off of a scientific model that is founded on an analogy. The latter is an important point because employing language in the context of a model that explicitly points to the analogy underlying the model reinforces the strength of that analogy in our thinking about the subject of the model.

5.4 Scientific Models Seen as Metaphors

The question is, on the one hand, what makes metaphors such appealing tools of thought, and on the other, why models are fruitfully compared to metaphors. I now single out the features of models discussed in association with the claim that scientific models are metaphors. The points that follow presuppose that the metaphors in question suggest analogies and that something like the cognitive claim attached to the interaction view ("new insight through metaphor") holds.

5.4.1 Familiarity and understanding

Often models and metaphors exploit the strategy of understanding something in terms of something else that is better understood and more familiar; they exploit the analogy relationship suggested by a metaphor or explored in a model. Of course, being familiar does not equate with being understood, but familiarity can be a factor in understanding. This is also not to suggest that understanding can be reduced to the use of analogy, but having organized information in one domain (source) of exploration satisfactorily can help to make connections to and achieve the same in another domain (target). The aim is to apply the same pattern in the target domain as in the source domain, with the same assumptions of structural relationships in both the source and the target domains. For instance, to think of the energy generation process in quasars in terms of energy generation in binary stars is helpful because it was by studying binary star systems that the importance of accretion of mass as a power source was first recognized. Moreover, turning the gravitational energy into the "internal" energy of a system is perhaps the only way to account for the enormous energies that must be present in quasars. The proposed conversion process of gravitational energy is, in turn, inspired by disks in planet or star formation. Piecing together these ideas based on analogies to already better-analyzed empirical phenomena paved the way to the formulation of the accretion disk model that is constitutive in explaining energy present in quasars and radio galaxies.[8]

5.4.2 Material for exploration

Models and metaphors can be hypothetical and exploratory. Besides a positive analogy that may have given rise to the formulation of a model or metaphor, there are negative and neutral analogies that can be explored (see chapter 3, section 3.3). This exploration furthers creative insight, as the interaction view proposes, because sometimes negative and neutral analogies offer a pool of ideas of what can be tested about the target domain. Metaphorical models nevertheless have to stand up to empirical reality, which is why Hesse (1966, p. 170) talks about "clearer truth criteria than for poetic metaphors" and "the (perhaps unattainable) aim to find a 'perfect metaphor'"—that is, a perfect description, one that provides an empirically adequate description of a phenomenon. An example for metaphorical exploration is artificial neural networks as used in computing for pattern recognition. Digital computers are serial processors and good at serial tasks such as counting or adding. They are less good at tasks that require the processing of a multitude of diverse items of information, tasks such as vision (a multitude of colors and shapes and so on) or speech recognition (a multitude of sounds) at which the human brain excels.

The example of the brain demonstrates how to cope with such tasks through many simple processing elements that work in parallel and "share the job." This makes the system tolerant to errors; in such a parallel distributed processing system, a single neuron going wrong has no great effect. The idea of artificial neural networks was therefore to transfer the idea of parallel processing to the computer in order to take advantage of the assumed processing features of the brain. Moreover, the assumption that learning occurs in the brain when modifications are made to the effective coupling between one cell and another at a synaptic junction is simulated in artificial systems through positive or negative reinforcement of connections. Artificial neural networks produce impressive results in pattern recognition, even though there remain considerable negative analogies between them and the human brain. Not only do the number of connections differ hugely from the brain, but the nodes in artificial neural networks are highly simplified in comparison to neurons in the brain. Explaining the neural network metaphor involves becoming aware of its appropriate applications as well as its limits.[9]

5.4.3 Coping with negative analogies

Metaphors, analyzed as being connected to analogies, usually involve the statement of negative analogies; these do not tend to hinder the use of the metaphor, however. Scientific models, in contrast, require attention to so-called negative analogies. Even though models claim no more than to be partial descriptions, in order to use them efficiently, their users need to be aware of those descriptions that do not apply. Negative analogies are those aspects of a phenomenon that are either not described by a model or that are not described correctly in the sense that they are not in agreement with empirical data about the phenomenon. Knowing which aspects of a phenomenon a model does not address is part of the model. As shown, an artificial neural network does not simulate the structure of the human brain in every respect, but we need to know in what respects it does. Not spelling out disanalogies explicitly in a model can have detrimental effects. Some metaphors, especially if even the positive analogy is questionable, can be positively misleading—for example, the common interpretation of entropy as a measure of disorder. Consider the example of a partitioned box of which one half contains a gas and the other is empty. When the partition is removed, the gas spreads over both halves of the box. This constitutes an increase of entropy because it is extremely unlikely that all gas molecules will simultaneously occupy just one half of the box spontaneously. It is not clear why the second situation should be viewed as a state of "higher disorder" than the first; a more helpful way of modeling entropy is to talk about the number of available microstates per macrostate.

5.4.4 Types of metaphorical models

The preceding discussion allows us to distinguish different ways in which metaphor or metaphoricity enters into scientific modeling. Different combinations of models being metaphorical or generating metaphorical terminology, or both, seem to be possible. Certain models are considered metaphorical in the sense that a transfer from one domain to another has taken place, but where perhaps no specific metaphorical terminology is used in this model. In other cases, a structural relationship is hypothesized between two domains that warrants a transfer of one structure leading to the formulation of a model of a similar structure in the target domain. In addition, this transfer gives rise to metaphorical language use accompanying the use of the model—for example, *temperature* in simulated annealing or *noise* in observational astronomy.

Then there are models where the descriptive terminology employed is metaphorical, but the two domains involved in this metaphorical terminology are not related in structure. An example is gravitational lensing. All a gravitational lens has in common with an optical lens is that it bends a light ray. The *bending* of a light ray due to gravitation is, unlike the case of an optical lens, not interpreted in terms of the optical phenomenon of refraction, so the metaphor is not connected to any deeper structural analogy between gravitational and optical lenses. Finally, yet other scientific metaphors that can be found in popular culture are without impact on scientific modeling, which is why they can be disregarded for the current purposes. Examples are *litmus tests* in politics, a *critical mass* of participants needed before ideas can be *generated*, a military *nerve center*, learning by *osmosis*, being *tuned in* or *turned off*, somebody being an Elvis *clone*, and so on (Hutchinson and Willerton 1988).

What can be said in summary about the relationship between models and metaphors? Insightful metaphors are those that point to a useful analogy between phenomena of two different domains. The development of scientific models also often relies on analogies. Both the interpretation of models and that of metaphors frequently benefits from the analogies associated with them. Analogy deals with similar attributes, relations, or processes in different domains; exploited in models; and highlighted by metaphors. Note that neither metaphors nor models *are* analogies—they are descriptions. This raises the question, of course, whether at the cognitive level there is anything involved in the metaphor claim concerning scientific models that can *not* be reduced to analogy.

5.5 Conclusion

I began with an inspiring hypothesis of how scientific models shape our thought: "Men think in terms of models" (Deutsch 1951, p. 230). This hypothesis is mostly based on an intuition that the way we describe phenomena (for example, with models, with metaphors) reflects the way we think about them. Both Kuhn's notion of a paradigm and Hesse's metaphor approach are similarly suggestive. They rely on the assumption that there are certain ways of thinking about phenomena. A second assumption is that people do not necessarily change their ways of thinking easily. This is one of Kuhn's big theses that resistance to a change of paradigm in the light of challenging scientific evidence can hinder scientific progress. Unfortunately, Kuhn's analysis of "paradigm" is ambiguously sketchy.[10] Hesse's analysis of models as metaphors indicates how it is an act of creativity first to adopt a new metaphor for something and then to learn to think about a phenomenon in a way that is suggested by the metaphor. According to Hesse, this involves a meaning shift. I highlighted some of the limitations of this metaphor approach, which is why I am tempted to conclude that the claim that models are metaphors is itself a metaphor.

Notes

1. Note that my treatment of the notion of a paradigm is not intended to be comprehensive. For a more thorough treatment, see Hoyningen-Huene (1989) 1993, chapter 4.

2. For instance, Kuhn (2000, pp. 285–286), in his interview with A. Baltas, K. Gavroglu, and V. Kindi, warmly remembers Mary Hesse's (1963b) positive review of his *Structure of Scientific Revolutions.*

3. For a review of the work on metaphor and models at that time and before, see Leatherdale 1974.

4. Black ([1979] 1993, p. 30) later contends: "I am now impressed, as I was insufficiently when composing *Metaphor* [Black 1954], by the tight connections between the notions of models and metaphors. Every implication-complex supported by a metaphor's secondary subject, I now think, is a *model* of the ascriptions imputed to the primary subject: Every metaphor is the tip of a submerged model."

5. This is only one view on metaphor, albeit the view that has been influential in philosophy of science and in thinking about scientific models. The topic of metaphor has also been extensively and controversially addressed in the philosophy of language (for example, Davidson [1978] 1984; and Searle 1979) as well as in cognitive linguistics (for example, Kittay 1987; and Langacker 1987; Lakoff and Johnson 1980 and 1999; and Lakoff 1993), but not with the view of being applicable to scientific models.

6. This is a claim strongly contested by Davidson ([1978] 1984).

7. This claim put forward very carefully is later reaffirmed: "I still wish to contend that some metaphors enable us to see aspects of reality that the metaphor's production helps to constitute" (Black [1979] 1993, p. 38).

8. For other examples from astronomy, see Cornelis 2000.

9. For more details on this example, see Bailer-Jones and Bailer-Jones 2002.

10. Shapere (1964) has provided a detailed analysis arguing that Kuhn's claims about paradigms cannot exclusively be extracted from his historical analyses. This leaves us with the general issue of gaining access to a paradigm. Shapere rightly points out that Kuhn is sometimes cagey and sometimes contradictory about this issue. Shapere (ibid., p. 386) comments: "In Kuhn's view . . . the fact that paradigms cannot be described adequately in words does not hinder us from recognizing them: they are open to 'direct inspection' (p. 44), and historians can 'agree in their *identification* of a paradigm without agreeing on, or even attempting to produce, a full *interpretation* or *rationalisation* of it' (p. 44). Yet the feasibility of a historical inquiry concerning paradigms is exactly what is brought into question by the scope of the term 'paradigm' and the inaccessibility of particular paradigms to verbal formulation."

References

Achinstein, P. 1968. *Concepts of Science.* Baltimore, Maryland: Johns Hopkins University Press.

Aronson, J. L., R. Harré, and E. C. Way. 1995. *Realism Rescued: How Scientific Progress Is Possible.* Chicago: Open Court.

Bailer-Jones, D. M. 2000. Scientific models as metaphors. In F. Hallyn, ed., *Metaphor and Analogy in the Sciences.* Dordrecht: Kluwer Academic Publishers, pp. 181–198.

———. 2002. Models, metaphors, and analogies. In P. Machamer and M. Silberstein, eds., *Blackwell Guide to Philosophy of Science.* Oxford: Blackwell, pp. 108–127.

Bailer-Jones, D. M., and C. A. L. Bailer-Jones. 2002. Modelling data: Analogies in neural networks, simulated annealing, and genetic algorithms. In L. Magnani and N. Nersessian, eds., *Model-Based Reasoning: Science, Technology, Values.* New York: Kluwer Academic/Plenum Publishers, pp. 147–165.

Bhushan, N., and S. Rosenfeld. 1995. Metaphorical models in chemistry. *Journal of Chemical Education* 72: 578–582.

Black, M. 1954. Metaphor. *Proceedings of the Aristotelian Society* 55: 273–294.

———. 1962. *Models and Metaphors.* Ithaca, New York: Cornell University Press.

———. [1979] 1993. More about metaphor. In A. Ortony, ed., *Metaphor and Thought.* Cambridge: Cambridge University Press, pp. 19–41.

Boyd, R. 1993. Metaphor and theory change: What is "metaphor" a metaphor for? In A. Ortony, ed., *Metaphor and Thought.* Cambridge: Cambridge University Press, pp. 481–532.

Bradie, M. 1998. Models and metaphors in science: The metaphorical turn. *Protosociology* 12: 305–318.

———. 1999. Science and metaphor. *Biology and Philosophy* 14: 159–166.

Cornelis, G. C. 2000. Analogical reasoning in modern cosmological thinking. In F. Hallyn, ed., *Metaphor and Analogy in the Sciences.* Dordrecht: Kluwer Academic Publishers, pp. 165–180.

Davidson, D. [1978] 1984. What metaphors mean. In *Inquiries into Truth and Interpretation.* Oxford: Clarendon Press, pp. 245–264.

Deutsch, K. W. 1951. Mechanism, organism, and society: Some models in natural and social science. *Philosophy of Science* 18: 230–252.

Gentner, D., and A. L. Stevens, eds. 1983. *Mental Models.* Hillsdale, New Jersey: Lawrence Erlbaum Associates.

Hanson, N. R. 1958. *Patterns of Discovery.* Cambridge: Cambridge University Press.

Harré, R. 1960. Metaphor, model, and mechanism. *Proceedings of the Aristotelian Society* 60: 101–122.

———. 1970. *The Principles of Scientific Thinking.* London: Macmillan.

———. 1988. Where models and analogies really count. *International Studies in the Philosophy of Science* 2: 118–133.

Hesse, M. 1953. Models in physics. *British Journal for the Philosophy of Science* 4: 198–214.

———. 1963a. *Models and Analogies in Science.* London: Sheed and Ward.

———. 1963b. Review of Thomas Kuhn, *The Structure of Scientific Revolutions. Isis* 54: 286–287.

———. 1966. *Models and Analogies in Science.* Notre Dame, Indiana: University of Notre Dame Press.

———. 1983. The cognitive claims of metaphor. In J. P. van Noppen, ed., *Metaphor and Religion.* Brussels: Study Series of the vrije Universiteit Brussel, pp. 27–45.

Hoyningen-Huene, P. [1989] 1993. *Reconstructing Scientific Revolutions.* Chicago: University of Chicago Press.

Hutchinson, B., and C. Willerton. 1988. Slanging with science. *Journal of Chemical Education* 65: 1048–1049.

Hutten, E. H. 1954. The role of models in physics. *British Journal for the Philosophy of Science* 4: 284–301.

Johnson-Laird, P. N. 1983. *Mental Models.* Cambridge: Cambridge University Press.

Kittay, E. F. 1987. *Metaphor: Its Cognitive Force and Linguistic Structure.* Oxford: Clarendon.

Kuhn, T. S. [1962] 1970. *The Structure of Scientific Revolutions.* Chicago: University of Chicago Press.

———. [1974] 1977. Second thoughts on paradigms. In *The Essential Tension.* Chicago: University of Chicago Press, pp. 293–319.

———. [1979] 1993. Metaphor in science. In A. Ortony, ed., *Metaphor and Thought.* Cambridge: Cambridge University Press, pp. 533–542.

———. 2000. *The Road since Structure: Philosophical Essays, 1970–1993.* Edited by J. Conant and J. Haugeland. Chicago: University of Chicago Press.

Lakoff, G. 1993. The contemporary theory of metaphor. In A. Ortony, ed., *Metaphor and Thought.* Cambridge: Cambridge University Press, pp. 202–251.

———, and M. Johnson. 1980. *Metaphors We Live By.* Chicago: University of Chicago Press.

———. 1999. *Philosophy in the Flesh: The Embodied Mind and Its Challenge to Western Thought.* New York: HarperCollins Publishers.

Langacker, R. W. 1987. *Foundation of Cognitive Grammar, Vol. 1: Theoretical Prerequisites.* Stanford, California: Stanford University Press.

Leatherdale, W. H. 1974. *The Role of Analogy, Model, and Metaphor in Science.* Amsterdam: North Holland.

Machamer, P. 2000. The nature of metaphor and scientific descriptions. In F. Hallyn, ed., *Metaphor and Analogy in the Sciences.* Dordrecht: Kluwer Academic Publishers, pp. 35–52.

Martin, J., and R. Harré. 1982. Metaphor in science. In D. S. Miall, ed., *Metaphor.* Sussex: Harvester Press, pp. 89–105.

Mastermann, M. 1970. The nature of a paradigm. In I. Lakatos and A. Musgrave, eds., *Criticism and the Growth of Knowledge.* Cambridge: Cambridge University Press, pp. 59–89.

Meyer, H. 1951. On the heuristic value of scientific models. *Philosophy of Science* 18: 111–123.

Miller, A. 1996. *Insights of Genius.* New York: Springer-Verlag.

Nersessian, N. 2003. Kuhn, conceptual change, and cognitive science. In T. Nickles, ed., *Thomas Kuhn.* Cambridge: Cambridge University Press, pp. 178–211.

Nickles, T. 2003. Normal science: From logic to case-based and model-based reasoning. In T. Nickles, ed., *Thomas Kuhn.* Cambridge: Cambridge University Press, pp. 142–177.

Paton, R. C. 1992. Towards a metaphorical biology. *Biology and Philosophy* 7: 279–294.

Richards, I. A. 1936. *The Philosophy of Rhetoric.* New York: Oxford University Press.

Rumelhart, D. E. [1979] 1993. Some problems with the notion of literal meanings. In A. Ortony, ed., *Metaphor and Thought.* Cambridge: Cambridge University Press, pp. 71–82.

Searle, J. 1979. *Expression and Meaning.* Cambridge: Cambridge University Press.

Shapere, D. 1964. The structure of scientific revolutions. *The Philosophical Review* 73: 383–394.

Wittgenstein, L. 1953. *Philosophical Investigations.* German-English edition. Translated by G. E. M. Anscombe. Oxford: Blackwell.

6 THE SEMANTIC VIEW AND THE STUDY OF SCIENTIFIC PRACTICE

I SPENT SOME TIME explaining, in chapter 4, how theories were analyzed to the detriment of the study of models. Then I emphasized the need for models in science, highlighting their role in discovery and creativity, in chapter 5. The movement of models toward the center of philosophical attention has continued since, as I illustrate in this chapter and also in chapter 8. Yet moving models to the center and out of the shadows obviously also changes the relationship *between* models and theories. The aim of this chapter is to elaborate where theory stands in this new picture. Although I already pointed to the scope of models in a number of places, especially in chapter 1, it is time to spell out the role and features of theory regarding models.

In chapter 4, I considered how theories were understood in the days of Logical Empiricism, an understanding of theory that dominated the philosophical scene at least until the 1950s. This view of theories is today often referred to as the "Received View," or the "Syntactic View," or sometimes the "Statement View." In the 1960s, the concern with scientific discovery and scientific change increased and resulted, for instance, in Kuhn's concept of paradigm change and in the metaphor approach to scientific models (chapter 5), but this was not the only approach to the issue of models. There also developed a more formal reaction to the Syntactic View of scientific theories. This is the so-called Semantic View of theories, which derives much of its inspiration from mathematical model-theory.[1] In this approach to models and theories, which I present in section 6.1, models become more central tools for describing the world, although a concern remains to what extent this approach really applies to models as encountered in scientific practice.

Studying the scientific practice was certainly one reason why models were beginning to be considered central to doing science.[2] Although the approach to models from scientific practice is opposed to the formal style and arguments of the Semantic View, the overall effect was similar: Theories got increasingly demoted to the backstage, while models—regarding the attention given to them and the claims made about them—conquered center stage. In section 6.2, I consider the argument from scientific practice quite generally and its applicability to the study of scientific models. Section 6.3 reviews Margaret Morrison's claim that scientific models are mediators between theory and the world and that they are autonomous agents. Section 6.4 recapitulates Nancy Cartwright's position on models as she developed it over the years. The account of Cartwright's views provides the basis for a model-theory distinction that I endorse in section 6.5. Section 6.6 uses the classical example of the pendulum to illustrate the points made about theories and models in section 6.5, and section 6.7 contains the conclusions of the chapter. My resulting proposal of theories being abstract and being applicable to phenomena only via models stands in the overall tradition of Morrison and Cartwright, even if I have some reservations concerning Morrison's characterization of models as autonomous agents and Cartwright's (sometime) suggestion that theories are true of models. I agree, however, with Cartwright's claim that theories are abstract and provide a straightforward account of what it means for a theory (or some models) to be abstract.

6.1 The Semantic View of Theories

In a recent review, Frederick Suppe (2000, p. S103) lists the reasons for the failure of the Received View. Some of those he lists are: the observational-theoretical distinction was untenable; correspondence rules were a "heterogeneous confusion"—that is, were taken to be too many different things; theories are not axiomatic systems; and symbolic logic is an inappropriate formalism. Interestingly, even though the Semantic View comprises a formal approach, arguments in its favor make reference to scientific practice. Suppe (ibid., p. S105) states: "For [Patrick] Suppes and myself an important motivation was making sense of personal experimental-scientist experiences we could not reconcile with the Received View."

Correspondingly, Patrick Suppes (1961, p. 165) seems to suggest that models in science and models in model-theory can be identified: "I claim that the concept of model in the sense of Tarski [a model of T is a possible realization in which all valid sentences of a theory T are satisfied] may be used without distortion and as a fundamental concept in all of the disciplines from which the above quotations are drawn [mathematical logic, spectroscopy, atomic physics, statistical mechanics, game theory, sociology, learning theory, probability theory]. In this sense I would assert that the meaning of the concept of model is the same in mathematics and in the empirical sciences."

Suppes then goes on to distinguish between the use of models in science and the meaning of the concept of a model as employed in well-defined technical contexts.[3] The Semantic View derives its concept of a model from such a technical context—namely, from mathematical model theory. Proponents of the Semantic View, despite acknowledging some differences between models in scientific practice and models in model theory (Suppes 1962, p. 252; and van Fraassen 1980, p. 44), maintain that the logical concept of a model *can* be applied to scientific models and that "the usages of 'model' in metamathematics and in the sciences are not as far apart as has sometimes been said" (van Fraassen 1980, p. 44). Ronald Giere (1999, p. 44), in turn, attributes considerable value to the concept of models as used in the Semantic View, in formal logic and foundations of mathematics, but doubts that this conception is suitable for understanding models as used in scientific practice.

Whatever the merits of the Semantic View, one thing that cannot be denied is that it specifically addresses the relationship between theories and models (cf. Suppe 2000, p. S109).[4] In mathematical logic, a model is a possible realization that satisfies the theory (that is, renders it true). The model consists of a relational structure satisfied by the sentences of which the theory consists (Suppes 1961, p. 166). The important point is that the structure of the model is such that it does not lead to contradictions within the theory. Suppes (1962, p. 252) writes: "Roughly speaking, a model of a theory may be defined as a possible realization in which all valid sentences of the theory are satisfied, and a possible realization of a theory is an entity of the appropriate set-theoretical structure." The idea is that a theory is identified with a class of models. The models are "non-linguistic entities"—that is, they do not contain propositions but are structures of elements.

Suppe (1974, p. 222) portrays not just models but theories too as nonlinguistic structures and distinguishes between the theory and the description of the theory. Of course, the *description* of a theory is a linguistic entity. However, given that the theory itself is nonlinguistic, it is possible that the same theory can be described in different ways—that is, can have a number of different descriptions.[5] Suppe (ibid.) comments: "Theories are extralinguistic entities which can be described by their linguistic formulations. The propositions in a formulation of a theory thus provide true descriptions of the theory, and so the theory qualifies as a model for each of its formulations." So, according to Suppe, theories are models of the linguistic formulations of theories. After all, as Anjan Chakravartty (2001, p. 327) has pointed out: "One of the primary motivations for the model-theoretic approach has been to escape worries about how linguistic entities link up with the world"—that is, one of the motivations of the Semantic View was to get rid of correspondence rules, simply by formally not addressing the issue. One problem with correspondence rules is that they depend on a sharp theory-observation distinction. Their construction does not make sense when observation itself is also theory-laden.

Of course, it is an advantage to be able to admit that a theory can have different formulations. After all, there exist examples for this—for instance, the Lagrangian and the Hamiltonian formulations of classical mechanics. The problem with models, or theories, that are nonlinguistic is that they do not *tell* us anything about the world, and this is in contrast to my premise that models tell us about phenomena that arise in the world (chapters 1 and 7).[6] Chakravartty (2001) makes roughly this point when he highlights the difficulties of being a scientific realist, while maintaining the Semantic View. One might, of course, suggest that it is easier to compare two nonlinguistic entities (a phenomenon and a model, according to the Semantic View), but this overlooks the fact that these nonlinguistic entities are intrinsically not very similar. One is concrete in that it is a fact or event in the real, empirical world and the other is not. A phenomenon has empirical properties, whereas a model that is a structure does not. If a phenomenon *has* a structure that could be compared to a model, then this structure is not particularly evident, and empirically not easily accessible. So comparing a model with a phenomenon clearly presents a not very trivial task, and probably one no easier than trying to deal with correspondence rules in a propositional approach (cf. ibid., p. 330).

As Chakravartty also points out, his considerations do not really depend on whether the position to be defended is a full-blown realism (although this is the position he considers). There may be other and weaker forms of how models relate to the world (see chapter 8 on representation), but the expectation is in any case that a model tells us *something* about the real world. For Chakravartty (ibid., p. 331), realism and the Semantic View do not go together, because "[a] model can tell us about the nature of reality only if we are willing to assert that some aspect(s) of the model has a counterpart in reality. . . . Scientific realism cannot be entertained unless we are willing to associate models with linguistic expressions (such as mathematical formulae) and interpret such expressions in terms of correspondence with the world."

If a model itself is nonlinguistic and we have to have a "description" of the model in order to decide how well the model relates to the world, then it is extremely tempting to conclude that the nonlinguistic model (as conceived in the Semantic View) is not a model as it stands. This implies that ultimately the Semantic View still fails to get rid of the problem-ridden statements of the Statement View. Models as conceived in the Semantic View are not the same kind of entities as models that tell us what the world is like. I am, however, concerned with models that tell us what the world is like, and these may or may not be "descriptions of theories," depending on the examples one considers.

As far as the claim is concerned that the Semantic View is in agreement with scientific practice, some philosophers have reconstructed scientific theories in accordance with the specifications of the Semantic View. The origins of the Semantic View of theories go back to a book by Evert Willem Beth called *Natuurphilosophie* that was published in 1948 (cf. van Fraassen 1987, p. 105). Beth (1949) applied his approach to Newtonian and quantum mechanics, and the Semantic View has also been applied to classical particle mechanics (McKinsey and Suppes 1953a and 1953b), to quantum mechanics (van Fraassen 1991), to evolutionary theory and population genetics (Beatty 1980 and 1981), and other theories (and see Craver 2002 for more references). However, the possibility to reconstruct certain existing scientific theories along the lines of the Semantic View is no proof that the Semantic View relates to scientific practice. While providing a tidy formal account is an admirable goal, the chances are that such an account does not do justice to the great variety of scientific models encountered in scientific practice.

6.2 Scientific Practice and Case Studies

From the 1960s, the historical study of science received increasing attention in philosophical contexts. Thomas Kuhn's *Structure of Scientific Revolutions*, first published in 1962, presented a milestone in this development (Rouse 2003). With this also came the criticism that the philosophical reconstruction of science and the actual practice of science are sometimes separated by a wide abyss. Correspondingly, arguments from scientific practice in favor of one or the other philosophical position became increasingly important. Case studies from all walks of science have become prominent tools of the philosophical discourse about science. In this section, I briefly consider the status of such arguments, as this type or argument is prevalent in many current discussions about scientific models.

What is the status of arguments from scientific practice in the philosophy of science? The one dilemma of any claims about the world based on case studies concerns how the case studies are chosen (Pitt 2001). There is, after all, likely to be a bias in selecting cases such that they support a certain view—a selection effect depending on the philosophical point one wants to make. Even if this were not so, the number of cases looked at is always going to be limited, and it is not clear how such a limited number of cases would warrant some general conclusion. The assumption behind this criticism is that philosophical claims are, almost by their nature, general. So whenever case studies are employed to argue that models *are* a certain way (true, false, idealized, simplified, or whatever, concurring with the Semantic View or not), then the cases studied are employed as arguments that something holds in a particular example case.

At the same time, there is the temptation to derive some more general claim from this. This kind of suggestion of generality will always leave copious space for the sceptic: Are models really like that, even if one considered *all* the examples? Sometimes models may look more like this, and sometimes more like that. It is interesting that, although we can never ultimately dispel the sceptic's argument, we constantly seem to try and generalize and put into a pattern what we find. This is what we seem to be set up to do as philosophers. I do not necessarily want to suggest that philosophers draw hasty conclusions from a biased sample of examples. Yet no matter how good and how representative examples are, they are only examples and not a complete set of cases, rather like in the problem of induction.

Another way to avoid the dilemma of how to do the philosophy of science is to acknowledge that there is more than one way to interpret science (Burian 2001). It is only possible to make generalizations within certain limited contexts. In other words, there simply is no one general pattern to be found in the world. This also means that the whole scope of the project of philosophy is more limited. Correspondingly, R. M. Burian (ibid., p. 401) has talked about a "reduction in the ambitions of that discipline." However, if we go to the extreme of never daring to generalize what we discover out there in the world and how we interpret it, then philosophy loses its subject. It is part of thinking about the world to try out different patterns and generalizations that may fit that world, so what is needed is a methodology that makes case studies profitable for philosophical purposes (Pinnick and Gale 2000). What case studies give us is a more detailed knowledge than we would have if we stayed at the quite general level of considering issues. This is why case studies, or particulars, from science can prevent us from grave errors in our interpretation of science. If we then go beyond the case study in our philosophizing, we must become speculative. Yet to the same degree to which we become speculative (and philosophically interesting), we lose the safety of the empirical foundation provided by the case study, precisely because it was only a particular case study.

To give a sense of the kind of findings that may result from a case study of a scientific model, let me sketch a case study that I conducted and the insights that it suggests. The case study is about modeling extended extragalactic radio sources (EERSs), astronomical phenomena outside our Galaxy, at enormous distances from the Earth and observable at radio wavelengths. This is an example of modeling *in progress*, modeling as part of a continuing effort to build an account of a phenomenon (Bailer-Jones 2000; see also Bailer-Jones 1999). Access to astronomical objects, such as EERSs, is quite restricted—they cannot be experimented upon—and much depends on observational and technical skill. Correspondingly, the modeling efforts regarding such astronomical objects meet comparatively few empirical constraints. The task confronted is one of "*accounting for a phenomenon in its entirety while having to confront a multitude of empirical features*" (Bailer-Jones 2000, p. 50; italics in original).

When examining EERSs, researchers encounter features and processes of the phenomenon that cannot be explained and accounted for—that is, there are *gaps* in the account of the phenomenon. The first task then is to identify important questions, "things that need to be modelled." All along, modeling EERSs also has a strong visual component in that the observed data are displayed in the form of radio maps that give a sense of what these objects "look like." The maps are based on observations in the radio spectrum—that is, not the optical—so the objects could never be "seen" in the way they are drawn on maps. So the purpose of maps is visualization for theoretical purposes, not to be a "picture" of the object in any ordinary sense. Visualizations of theoretical processes also play a role in ordinary research contexts and often seem to serve as *reifications* of suggested theoretical processes (Bailer-Jones 2002).

Another important point emerging from this case study is that the modeling process is split up. Many submodels are construed addressing individual gaps in the account of the phenomenon. Portioning the problem-solving task in this way makes the individual tasks much more manageable, and scientists with their specific expertise can work on individual submodels, while remaining comparatively unconcerned with the issues arising in other submodels. Despite splitting up features of the phenomenon for the purposes of modeling, the submodels still remain constituents of an overall model and are expected to be sufficiently consistent with each other to be parts of one overall account. To re-create the fact that the submodels all address one and the same phenomenon, they have to be *embeddable* into an overall model. This process of embedding often reveals shortcomings of the models in that they do not fit together perfectly. Other shortcomings encountered in submodels are inaccuracies, simplifications, and even downright disagreement with empirical data (Bailer-Jones 2000); some models are "false" and yet productive of insight (Wimsatt 1987). The point of a case study like this is that the claims about modeling (that are merely sketched above) are substantiated: they are supported in detail by example; evidence is provided that there exists at least one instance of this kind. This gives confidence that the philosophical account given is correct at least for the example studied. It is no basis for confidence in what happens in other instances, but it provides a guideline for what should be examined in other examples and what should be considered philosophically.

6.3 Models as Autonomous Agents

Relying on case studies for philosophical claims has been a popular line to take in recent studies of modeling. The claim that models are mediators between theory and the world has arisen in this research context. This recent slogan of "models as mediators" appeals to many, and coeditors Mary Morgan and Margaret Morrison (1999) have put forward the position in their volume that carries the very title *Models as Mediators.*[7] The book contains mostly case studies that fit into the general framework of models as mediators. Again, this is a proposal of how models, theories, and the world relate to each other (see also chapter 7, especially section 7.5).

One claim about how models and theories relate is that models are not derived from theories. This is the claim of models being autonomous agents (Morrison 1998 and 1999). Models are said to be partially independent of theory and of data but involve both (Morrison and Morgan 1999, p. 11). Moreover, their partial independence from either theory or data is further endorsed by the suggestion that "other elements" are involved in their construction (ibid., p. 15; for examples, see Boumans 1999; cf. Cartrwright's toolbox approach, section 6.4.3). Providing the example of the pendulum (see section 6.6), Morrison and Morgan (ibid., p. 16) illustrate that models do not exclusively draw from theory for their construction: "[T]heory does not provide us with an algorithm from which the model is constructed and by which all modelling decisions are determined." Given how I have introduced the concept of a model in chapter 1, this claim sounds almost self-evident, but it does not in the context of the Semantic View, where the theory is comprised of models. However, while there exists partial independence, models also *need* to be related to theory and to the world, although sometimes more to one than to the other (ibid., pp. 25ff.).

I find talk of autonomy of models misleading. There is no denying that there always exists *some* relationship between a model and some theory from which the model draws, and between a model and the phenomenon of which it is a model. In short, there always exist constraints for the relationship between model and theory and model and phenomenon. Even if these constraints turn out to be particular—that is, if they were not the same for all models—there would still be constraints for each individual instance of model-phenomenon relationships. What the relationship is between a model and a phenomenon (or between a model and a theory) may partially (although not fundamentally, I think) vary from model to model. Of course, models are "autonomous" with respect to theories in that they do a different job from theories (cf. section 6.5), and for doing this job, models may use more or less theory, depending on the phenomenon modeled and the function of the model. Models are also, in some trivial sense, autonomous with respect to the world, because the model does not constrain the part of the world that is modeled. The model can be "however it likes" (good, bad, appropriate, suitable for a given purpose, and so on), and the part of the world that is modeled is not going to be affected as a result of this. (Of course, our interpretation of the world may change as a result of a model, cf. chapter 7, section 7.3). This, however, cannot mean that the development of a model is entirely independent of the features and properties of the phenomenon modeled. Of course, one needs to ask: How much accuracy is required in a specific case? How many obviously false assumptions are tolerated in a certain model? (For more on this, see chapter 8, section 8.3.) The terminology of "autonomy" clearly implies that there is an agent, especially as Morrison talks about "autonomous agents," yet it is clearly not the case that a model can be anything like a human agent who "decides" to "act" in some autonomous manner. In contrast, it is people who may decide how closely a model needs to relate to the world, and how tightly empirical and theoretical constraints have to be adhered to in a particular instance. Perhaps this great diversity of degrees of "empirical adequacy" and degrees of theory involvement is the point Morrison wants to communicate, but if this is so, then I find the metaphor of models as autonomous agents infelicitous.

6.4 Cartwright on Models

I discuss Nancy Cartwright's work on models in some detail because she has worked out a notion of theory which I take as a basis for characterizing the relationship between models and theories in the next section. Models have their role to play in the larger picture of Cartwright's philosophy of science. They are a component facilitating the explication of how science works. There are many queries one can put forward in response to Cartwright's account, but also many inspired ideas in it. The following sections show how Cartwright's understanding of models has evolved over the years and how, according to her, models relate to theory.

6.4.1 Models as fictions

The first manifestation of Cartwright's views on scientific models is her simulacrum account of explanation (Cartwright 1983). According to this account, explanatory power does not count as an argument for the factual truth of theories or models. A model may explain a phenomenon and yet not have any claim to truth by virtue of this. Instead, "[t]o explain a phenomenon is to find a model that fits it into the basic framework of the theory and that allows us to derive analogues for the messy and complicated phenomenological laws which are true of it" (ibid., p. 152). As the title of her book proclaims, Cartwright argues that the laws of nature can lie, thus the theoretical framework on which the model may be based does not warrant the truth of the model. Models are prepared specifically that fundamental laws can feature in them. She writes: "For the kind of antecedent situations that fall under the fundamental laws are generally fictional situations of a model, prepared for the needs of the theory, and not the blousy situations of reality"(ibid., p. 160). Thus laws do not literally apply to the real situations (ibid., p. 161). It is only phenomenological laws that can be true of phenomena. Phenomenological laws are laws as they are observed in phenomena; they are not integrated into a background of theories; they are formulated ad hoc.[8] Then, a model—which is based on certain theoretical laws—is merely an analogue to the messy phenomenological laws. Furthermore, different models have different purposes (ibid., p. 152), with the models having different emphases depending on their purpose. This sheds some further doubts on models being realistic. A model may serve a certain purpose well, without necessarily being realistic (ibid.)

Honoring such antirealistic tendencies of models, Cartwright proposes a "simulacrum account of explanation." She defines "simulacrum" in accordance with the *Oxford English Dictionary*, as "'something having merely the form or appearance of a certain thing, without possessing its substance or proper qualities'" (ibid., pp. 152–153). Things are "not literally" what their models say they are. Cartwright thus goes on to claim that "[a] model is a work of fiction. Some properties ascribed to objects in the model will be genuine properties of the objects modelled, but others will be merely properties of convenience" (ibid., p. 153). The aim of such "properties of convenience" of models is to make mathematical theory applicable to the objects that are modeled (ibid.). Models render theories, here assumed to be mathematical in character, applicable to phenomena, albeit with models having fictional status. This fictional status presumably has to do with the limitations of theories as not telling us how things "really" are: "I think that a model—a specially prepared, usually fictional description of the system under study—is employed whenever a mathematical theory is applied to reality, and I use the word 'model' deliberately to suggest the failure of exact correspondence" (ibid., pp. 158–159).

So models, on the one hand, fail to have an "exact correspondence" to the phenomena they represent and, on the other hand, they are needed for theories to establish some kind of relationship to reality. Cartwright writes (ibid., p. 159): "[O]n the simulacrum account, models are essential to theory. Without them there is just abstract mathematical structure, formulae with holes in them, bearing no relation to reality."[9] This characterization of theories (a) as abstract mathematical structure, (b) as formulae with holes in them, and (c) as not bearing any relation to reality is the core idea that provides the skeleton for the characterization of theory later in her article. The fables account of models, introduced in the next section, is an elaboration of this understanding of theories.

6.4.2 Models as fables

In an article some years later, Cartwright (1991) compares scientific models to fables. This is not about models being fictions. It is about the contrast between the abstract and the concrete. Fables have a moral that is abstract and they tell a concrete story that instantiates that moral, or "fits out" that moral. A moral of a fable may be "the weaker is prey to the stronger," and a way to "fit out" (Cartwright's formulation) this abstract claim is to tell the story of concrete events of the marten eating the grouse, the fox throttling the marten, and so on. Similarly, an abstract physical law, such as Newton's force law, $F=ma$, can be fitted out by different more concrete situations: a block being pulled by a rope across a flat surface, the displacement of a spring from the equilibrium position, the gravitational attraction between two masses. Thus Newton's law may be fitted out by "different stories of concrete events." Drawing from the analogy between models and fables, models are about concrete things; they are about concrete empirical phenomena.

The contrast between models and theories is not that theories are abstract and models are concrete. Rather, models are about concrete phenomena, whereas theories are not about concrete phenomena. If at all, theories are about concrete phenomena only in a very derivative sense. A second claim, beyond the distinction between the abstract and the concrete, has to do with something like existence. "Force," which is an abstract notion, does not manifest itself outside concrete empirical situations. Force is a factor in and contributing to empirical phenomena. Cartwright's everyday example for this relationship is "work": The abstract concept of work may be filled out by washing the dishes and writing a grant proposal, and this does not mean that a person washed the dishes and wrote a grant proposal, *and* worked—working does not constitute a separate activity—since working consists of just those activities. Cartwright (ibid., p. 65) explains: "*Force* —and various other abstract physics' terms as well—is not a concrete term in the way that a color predicate is. It is, rather, abstract, on the model of *working*, or *being weaker than*; and to say that it is abstract is to point out that it always piggy-backs on more concrete descriptions. In the case of *force*, the more concrete descriptions are ones that use the traditional mechanical concepts, such as *position*, *extension*, *motion*, and *mass*. Force then, on my account, is abstract relative to mechanics; and being abstract, it can only exist in particular mechanical models."[10]

Cartwright (ibid., p. 68) then infers that laws are true in models just as models are true in fables. However, this is not to say that the models need to be true of the world, just as the fables may not be true of the world. An abstract concept, such as force, may not be suitable to modeling *all* aspects of the world, perhaps only certain ones that are carefully constructed (in the laboratory, under exclusion of various other factors). It is for those situations where the abstract concept of force can be applied in a model that, according to Cartwright and in her terms, Newton's law is true of the model. I take models to be about phenomena in the world, so the important relation is not one of theories being true of models, as Cartwright sometimes suggests, but one about models being true (or something like that) of the world. In section 6.5, I endorse the suggestion that theories are abstract and that models are models of concrete phenomena of the empirical world.

Making this distinction obviously depends rather a lot on what "abstract" and "concrete" are taken to mean, but more on this below.

6.4.3 The toolbox of science

In Cartwright, Shomar, Suárez 1995, Cartwright revises her position on theories and models from her earlier statements. There she criticizes what she calls the "theory-dominated" view of science and sees herself as part of the "movement to undermine the domination of theory" (ibid., p. 138). It is models, rather than theories, that represent phenomena of the physical world (ibid., p. 139). Theories, in turn, are but one of the tools used in model construction. Other such tools are, for instance, scientific instruments or mathematical techniques. The change from her earlier view is that theories no longer represent the world via models, but rather that they do not represent the world. Correspondingly, Cartwright (ibid., p. 140) states: "I want to urge that fundamental theory represents nothing and there is nothing for it to represent. There are only real things and the real ways they behave. And these are represented by models, models constructed with the aid of all the knowledge and techniques and tricks and devices we have. Theory plays its own small important role here. But it is a tool like any other; and you can not build a house with a hammer alone."

Cartwright's general claims are illustrated by an example that her coauthors Towfic Shomar and Mauricio Suárez elaborate. The example is the 1934 model of superconductivity developed by Fritz and Heinz London, and the claim is that this model did not develop via the theory-driven strategies of approximation and idealization. A classic example of theory-driven modeling would be gradually modifying an equation to make it more realistic by adding correction terms —for example, when adding a linear term for mechanical friction to the equation of the simple harmonic oscillator, resulting in an equation for a damped linear oscillator. The superconductivity example is a case in point that not all scientific modeling is a process of de-idealization. Instead, there can be what may seem to be ad hoc adjustments to the theory that are not theory-driven but phenomenological. In sum, the argument is that there exists phenomenological model building in science that is perfectly valid yet independent of theory in its methods and aims (ibid., p. 148).[11] This scenario about the relation between theories and models is the extreme point of the demotion of theory: theory does not represent anything; it is but a tool in model construction. As I spell out next, this conclusion is also a direct consequence of the fables account.

6.4.4 Models in a dappled world

In accordance with Cartwright's earlier views that the laws of physics lie, it also holds for the dappled world that not everything that happens in the empirical world can be captured by the laws of physics, only those things for which there are models that match them. Cartwright (1998, p. 28) vividly illustrates this point with the example of a thousand dollar bill swept around in St. Stephen's Square. There exists no model of classical mechanics that is capable of describing this complex physical situation. In Cartwright's terms this means that classical mechanics is not universally applicable (in principle), and it means that the laws of mechanics do not determine this particular process. Instead, it may be necessary to switch to another area of physics for a description (for example, fluid dynamics). It may then be possible to have a model based on fluid dynamics that nearly enough captures what is going on with the thousand dollar bill. The point is again that any theory applies to the world only through its models. She writes: "Fluid dynamics can be both genuinely different from and genuinely irreducible to Newtonian mechanics. Yet both can be true at once because—to put it crudely—both are true only in systems sufficiently like their models, and their models are very different" (ibid., p. 29).

In the same way, quantum mechanics does not replace classical mechanics. Both theories make good predictions in certain real-world situations and are frequently employed in cooperation (ibid., p. 29). On Cartwright's (ibid., p. 30) account, the world is dappled, which is to say that classical and quantum mechanical accounts can hold at the same time. Turning the whole argument around, it is not only the case that laws apply within the limited range of a model, but also that models can serve as blueprints for "nomological machines," which provide the basis for arriving at a law (Cartwright 1997 and 1999a;

see chapter 3). Models tell us under which specific circumstances certain laws arise (Cartwright 1997, p. 293), in opposition to a Humean regularity view of laws that portrays laws as universal. So what is a nomological machine? Cartwright (1999a, p. 50) says: "It is a fixed (enough) arrangement of components, or factors, with stable (enough) capacities that in the right sort of stable (enough) environment will, with repeated operation, give rise to the kind of regular behaviour that we represent in our scientific laws."

Laws can be formulated under the specialized conditions created by the nomological machine. These conditions are mostly achieved by "shielding" (by controlling) the input into the machine such that anything is prevented from operating that might interfere with the machine functioning as prescribed. The result are ceteris paribus laws for the specific situation. Even probabilistic laws can be developed by means of nomological machines, so-called chance setups. In short, nomological machines produce the orderly and lawful outcome that is so much in contrast to the real, dappled world of the thousand dollar bill on St. Stephen's Square.

6.4.5 Representative and interpretative models

The limits of theory continue to be a topic for Cartwright (1999b, reprinted in Cartwright 1999a) in her article on the BCS model of superconductivity, where she puts her modeling view in the historical context of theory dominance in the philosophy of science. She not only rejects the Received View of scientific theories as axiom systems in formal languages because they lack expressive power (Cartwright 1999b, p. 241), she also discounts the Semantic View of theories, which considers models as constitutive of theories (ibid., p. 241). Instead, Cartwright adopts the view of models as mediators between theory and the real world (ibid., p. 242; Morrison and Morgan 1999; cf. section 6.3). The models that mediate between theory and world are *representative models* (Cartwright 1999b, p. 242), formerly *phenomenological models.* They represent the world not by being part of a theory (in contrast to *interpretative models*), although they may draw from theories. Cartwright (ibid., p. 242, and Cartwright 1999a, p. 180) takes representative models to be "models that we construct with the aid of theory to represent real arrangements and affairs that take place in the world—or could do so under the right circumstances."

Representative models can represent specific situations, and to do so they may go well beyond theory in the way they are built. This means that theory is not the only tool for model construction; others are scientific instruments, mathematical techniques, or the kind of laboratories, just as proposed in the toolbox approach. Thinking of theories, in turn, as abstract is already familiar from the notion of models as fables. Cartwright (1999b, p. 242) writes: "I want to argue that the fundamental principles of theories in physics do not represent what happens; rather, the theory gives purely abstract relations between abstract concepts: it tells us the "capacities" or "tendencies" of systems that fall under these concepts. No specific kind of behaviour is fixed until those systems are located in very specific kinds of situations."

Interpretative models, in turn, are models that are "laid out within the theory itself" (ibid., p. 243). Via bridge principles, the abstract terms of a theory can be made more concrete in an interpretative model. Interpretative models establish a link between abstract theory and model, whereas representative models establish the link between model and world (ibid., p. 262). So it would seem that representative models can be, but do not have to be, interpretative models. This is so insofar as interpretative models make abstract notions that feature in theories more concrete, and in that sense they can serve to represent certain situations that fall under the theory. Interpretative models have the function of representing certain theoretical situations, and these may or may not be similar to real situations.[12] Representative models, in turn, need not and do not have this interpretative function to "fit out" theories.[13]

A problem with the notion of representative models is that Cartwright does not elaborate the concept of representation, which she uses to say that theories do not represent the world and that representative models do (as in Cartwright, Shomar, and Suárez 1995). She does not want representation to be thought of as structural isomorphism (Cartwright 1999b, p. 261). According to her, the notion needs to be broader than one "based on some simple idea of picturing" (ibid., p. 262). It is a "loose notion of resemblance" that is instead suggested (ibid.). As Cartwright herself acknowledges, this is not much more than pointing to the problem of representation. (For more on the issue of representation, see chapter 8.)

6.5 Abstract Theories and Models of Concrete Phenomena

The point that Cartwright suggests, and which I want to endorse, is that theories are not the kind of entities which allow us to find out whether things are in the empirical world as the theory states. Let me go back to fables to illustrate this point. The moral of a fable, such as "the weaker is prey to the stronger," can, in this universal form, not be tested. There may be individual cases for which this moral or "theory" works, such as when the marten eats the grouse and the fox throttles the marten. In a dappled world, there may be *many* instances where the weaker is prey to the stronger, but it is exactly the *universality* with regard to which the moral or "theory" needs to be doubted, if we really live in a dappled world. There is absolutely no reason not to think that there could be other morals, or other theories, that provide a suitable base for a model. Just think of the race between the hedgehog and the hare. There, it is not the weaker, or in this case the slower, who loses out. The moral is that factors other than physical strength can play a role in winning a race.

Just as Cartwright asserts for the dappled world, both models have their justification. The world can be like in the fable of the hedgehog and the hare, or like in the fable of the grouse, the marten and the fox—hence like both fables. This simply depends on the individual situation that is in question. And indeed, as Cartwright claims elsewhere, theories can form "partnerships" where different theories are applied in different models when real phenomena cross over different areas of physics (Cartwright 1998, p. 34). The world offers a wide range of instances and cases, only some of which fall comfortably under one or the other moral. There are lessons about the world in "the weaker being prey to the stronger," as well as in "winning over sheer physical aptitude by means of wit." In view of the fables analogy and having to apply different morals to different empirical situations, it perhaps becomes easier to see how theories do not tell us anything *directly* about the world.

Cartwright characterizes theories as abstract in the context of the analogy between models and fables ("theories are like morals of fables"). I concentrate on one aspect only of theories being abstract. This aspect is that theories, being abstract, are not directly about empirical phenomena.[14] Abstractness is opposite to concreteness. The phenomena that are explored by modeling are *concrete* in the sense that they are, or have to do with, real things, real things with many properties—things such as stars, genes, electrons, chemical substances, and so on. Wanting to say that models are about concrete phenomena, while theories are not, brings with it still another problem, however. Often, the subject of models is a class of phenomena, rather than a specific individual phenomenon. Of most phenomena we can find many specimens in the world; these phenomena belong to the same class.[15] Modeling a star, there are many different individual stars that could serve as a prototype.[16]

One tries to model, however, not any odd specimen of a phenomenon, but a typical one. Often this involves imagining the object of consideration as having "average" or "typical" properties, and this "prototypical" object or phenomenon may not even exist in the real world. The point is that it could typically exist in just this way and that there probably exist many very much like it. So the prototype is selected or "distilled" from a class of objects. The prototype has all the properties of the real phenomenon; it is merely that the properties are selected such that they do not deviate from a "typical" case of the phenomenon. It is then this prototype that is addressed in the modeling effort. The assumption behind this process of prototype formation is nonetheless that the model is not only a model of the prototype, although there is a certain amount of deviation from the norm.

Correspondingly, modeling the human brain is not about modeling the brain of a specific person, but that, roughly, of all "typical," "normal" people. For my purposes, the prototype of a phenomenon still counts as concrete, because it has all the properties of the real phenomenon and *could* exist in just this manner. The target of the examination remains an empirical phenomenon, even if members of the class of phenomena that belong to a certain type deviate from the norm. A prototype-forming procedure is often needed to grasp and to define a phenomenon and to highlight what it is that one wants to model. The important point here is that despite prototype formation, the phenomenon is not in any way stripped of any of its properties.

Phenomena involve properties. Abstraction I take to be a process where properties are "taken away" from a phenomenon and ignored.[17] That which is abstract lacks certain properties that belong to any concrete phenomenon. To put it very crudely, something concrete becomes abstract when certain properties that belong to the "real thing" (and that make it concrete) are taken away from it.[18] Not all concepts, principles, or theories that are called abstract are abstract in the same way, but I think the notion of taking away some of those properties that make something concrete can still serve as a guideline. It is important to recognize that no theory is conceivable without the concrete instantiations from which the theory has been abstracted. We need to go through different example problems to understand how $F=ma$ is instantiated in different models. The theory is that which has been "distilled" from several concrete instantiations. In this sense, the abstract theory is not directly about concrete phenomena in the world. The properties that are missing in such an abstract formulation as $F=ma$ have to do with how the force makes itself noticed in different individual situations.

Think again of a block being pulled across a flat surface, or the displacement of a spring from the equilibrium position, or the gravitational attraction between two masses. It depends on the situation how the force is operative (deceleration due to friction, the repulsion of a spring, acceleration due to gravitation). Moreover, for application to each *concrete* situation one would have to establish what the body is like whose mass is operative in the physical system. Correspondingly, force, acceleration, and mass can be linked to different properties in different physical systems. Force, abstractly speaking, can be something that applies to an object or system, but force alone, without an object or a system, is not something about which we can say anything, nor know the properties of. To establish a theory, we need models that tell us how the theory is to be applied to the phenomenon or process modeled.[19]

Let me add a brief note on laws and theories. Some laws have the status of theories, but not all do. Some sciences may be hard-pressed to formulate theories or principles that are abstract enough to apply to a broad domain of issues of concrete instances, although an effort is often made. In other words, in the scenario I sketch there can be sciences that only employ models and do not have theories. However, there can also be laws that are merely generalizations of concrete instances—for example, "the melting point of lead is 327 degrees Centigrade," which is presumably true of all lead. This is not an abstract statement about lead. An abstract law would be one that told us, for instance, how to infer the melting point of quite different metals. (See chapter 7, section 7.1, for a brief discussion of how the melting point of lead is established.) For a law that simply states the melting point of lead, be it right or wrong—that is, for phenomenological laws—we do not need a model to apply it to the world. Such a law does not apply to a range of different instances from which it is abstracted; such a law applies only to one kind of instance. This makes the law not theoretical. Correspondingly, for such a law, testing its truth empirically is straightforward.

Theories can become general because they are abstract; they are free of the properties that are typical of the instances where the theory might apply, or the properties of prototypes. To model a phenomenon, abstract theory needs to be made more concrete, taking into account the specifications of the phenomenon that is modeled and inserting the boundary conditions of that phenomenon (or the prototype thereof). To see how the theory holds in a model, we need to fill in the concrete detail that is not part of the theory because, being abstract, the theory has been stripped precisely of those details.

To summarize, theory in science tells us much less what the world is like than models do. Theories are that to which we (sometimes) resort when we try to describe what the world is like by developing models. This is in overall accordance with Cartwright's position, although I have attempted to elaborate the point further. "Abstract," said of theories, means having been stripped of specific properties of concrete phenomena in order to apply them to more and different domains. Models, in turn, are about concrete phenomena (or prototypes thereof) that have all the properties that real things have. Theories are applied to real phenomena only via models—by filling in the properties of concrete phenomena. Being abstract and therefore not in an immediate sense about empirical phenomena does not, however, render theories worthless or unimportant. Theories and models have to prove themselves at different levels, models by matching empirical phenomena and theories by being applicable in models of a whole range of different phenomena (or prototypes thereof).

6.6 The Pendulum: A Classical Example

In my characterization of the relationship between models and theories, I have omitted an extremely important element so far: namely, the idealization that takes place in models or when theory is applied to models (see also Suárez 1999). Idealization is part of modeling. Needless to say, if idealization has taken place, then the idealized object or phenomenon is still expected to have a significant amount in common with the real phenomenon, or the idealized model is still expected to tell us something significant about the phenomenon ("represent it," cf. chapter 8, section 8.3). Conceptually it makes a difference whether it is the model that is idealized or what is taken to be the phenomenon modeled (McMullin 1985). Ernest McMullin has called the former *construct idealization* and the latter *causal idealization* (see again chapter 8, section 8.3).

Idealization is a process whereby existing properties of a phenomenon are changed deliberately to make modeling easier (cf. Cartwright 1989, pp. 185ff.; and Chakravartty 2001, p. 328). For instance, experimenting on Mendel's laws, one might need to decide how many peas of a generation come out wrinkled and how many smooth. Some peas may be somewhat wrinkled, although not very. For the model one has to decide whether certain peas count as wrinkled or smooth. So a semiwrinkled pea may, for the purposes of being modeled, have its properties "changed" into counting as a wrinkled pea, to fall into the prescribed categories. This is an instance of idealization. Similarly, modeling the ideal pendulum, we need to set up the situation such that Newton's force law can be applied. We assume that the string of the pendulum has no mass. Thus a real string with a mass has been "changed," for the purposes of modeling, into a string that has the property of having no mass. All the idealizations are instances whereby the properties of the phenomenon modeled are changed. In reality,

- the string is not weightless;
- the string is not inextensible;
- the string is not rigid;
- the mass of the pendulum bob is not located in one point.

So all these are idealizations carried out "constructing" an ideal pendulum. Not surprisingly, this kind of ideal pendulum is quite different from real pendulums, as they may be encountered in the world. Consequently, the model of the ideal pendulum is unlikely to be a very realistic representation of certain aspects of real pendulums. According to Anjan Chakravartty (2001, p. 329): "Idealization . . . cannot be adopted so straightforwardly by a realist analysis. Model assumptions here *contradict* what we take to be true of reality. Realism in this context will be carefully qualified at best." A. Bartels (2002), for the same reason, has described the ideal pendulum as fictional, and Ronald Giere (1988, p. 77) has pointed out that the laws of motion would have to be a lot more complex to provide an "exact description of even the simplest of physical phenomena."

Let us look at the example of the pendulum in a bit more detail to study the use of theory in this model. An ideal pendulum is a pendulum that consists of a particle of a certain mass, m (a "pointmass"), which is attached to one end of a weightless inextensible cord of length, l (for example, Frauenfelder and Huber 1966, p. 124). The task then is to determine the force with which the pendulum bob is pulled toward its equilibrium position—under the idealized circumstances. The force enters into the treatment of the pendulum in the shape of an abstract theory, Newton's second law $F=ma$. To think about the force in this particular case, it is necessary to take into account the specifics of this system, first of all the geometry of the system, involving the displacement angle, the length of the string, and the gravitational attraction of the Earth. The treatment of the ideal pendulum then usually continues by assuming that the displacement angle of the pendulum is small, because this allows us to replace the sine of that angle with that angle itself. (This last step is an instance of construct idealization. A change is made to the mathematical description of the phenomenon, not to the properties of the phenomenon itself.)

Obviously, physicists modeling pendulums are fully aware of the idealizations they have introduced into their model. It is for a good reason that they specifically talk about the "ideal" or "mathematical" pendulum when they refer to this particular model involving the specified idealizations. The idealizations are necessary to apply Newton's force law. Then other theories and considerations become necessary to make the ideal pendulum into a more real pendulum, to add corrections for the deviations from reality that have been introduced through idealization. Thus physicists consider the physical pendulum. This is supposed to be a model that is nearer to some real pendulums. This kind of pendulum is taken to be a rigid body of any arbitrary shape, pivoted about a fixed horizontal axis. In this case, the center of mass is treated as if it were the pendulum bob and the moment of inertia about the axis of rotation plays a role when calculating the restoring force.

It is then also possible to produce an equation that indicates which physical pendulum corresponds to an ideal pendulum of a different length—that is, one can talk about the *equivalent length* of a physical pendulum.[20] The pendulum can also be made "more real" in other ways. For instance, one can correct for the frictional forces of the air resistance, the buoyancy of the pendulum bob (the fact that the apparent weight of the bob is reduced by the weight of the displaced air), and the gravitational field of the Earth not being uniform, and so on (cf. Morrison 1999, pp. 48–51).[21] It is clear that, in order to make these corrections to the model, theories other than Newton's law in the case of gravitational attraction are employed and customized to the problem in hand. Of course, in this case they all fall under the reign of classical mechanics, but this need not be so. Morrison (1999, p. 51) comments: "We know the ways in which the model departs from the real pendulum, hence we know the ways in which the model needs to be corrected; but the *ability* to make those corrections results from the richness of the background theoretical structure." This example illustrates well the amount of customizing and introducing of specific details about a phenomenon to be modeled that goes into applying a theory to a phenomenon by means of a model.[22]

6.7 Conclusion

The goal of this chapter has been to reconsider how models and theories are related. I reviewed a number of responses to the understanding of theory that had developed in the tradition of Logical Empiricism (see chapter 4), the so-called Received or Syntactic View. According to this view, theories consist of propositions that are related to the world by means of correspondence rules. The first change, commencing in the 1950s, was that models were considered at all in the context of philosophy of science; the second change was that they were thought to be important (see also chapter 5)—to the extent that today some would claim that models are more important than theories. Models and theories need to be repositioned in current philosophy of science, and I retraced the path toward their new positions.

The Semantic View (section 6.1) took its lead from mathematical model theory and theories were interpreted as sets of models. I discussed two concerns with the Semantic View. One is that it portrays models as nonlinguistic entities ("structures"), and it is hard to see how models represent phenomena without some propositional element in them. The second concern is to what extent the Semantic View can capture and do justice to the role models play in scientific practice. Here, different claims can be heard, yet it is a general question of how to develop philosophical arguments drawing evidence from scientific practice. To what extent do philosophical accounts need to "fit" scientific practice or even take their lead from there? I discussed this general issue in section 6.2. Section 6.3 criticized the claim that models are mediators between theories and phenomena. Although this approach gets some things right about the position of models in the scientific enterprise, it is not always easy to specify what the claim involves. In section 6.4, I recapitulated Nancy Cartwright's views on models and theories over the years. I find her claims inspired and worth pursuing further, even if not always sufficiently elaborated in her own work. This is why, in section 6.5, I tried to go some way toward elaborating what it means for theories to be abstract and not directly applicable to empirical phenomena. In section 6.6, I illustrated these claims with the classical example of modeling the pendulum.

I now summarize the results of my examination. Models are positioned between theories and phenomena. They customize theory in a way that makes theory applicable to phenomena. So one *could* say that models "mediate" between theories and phenomena, but this claim has not quite the same connotations as in Morrison's work. Models are clearly much more than a join. On this, Morrison and I agree. They are central to how theory gets applied. But it is wrong to stylize models as autonomous agents, because they are subject to constraints both from the direction of theory and from the direction of the phenomenon modeled. What these constraints are may be negotiable in the individual case, but constraints they are. I subscribe to Cartwright's general line of characterizing theories as abstract. In the account I developed, this means that theories do not aim to capture the concrete properties of phenomena. Because many specific properties of phenomena are omitted in the formulation of theories, theories become widely applicable: to a whole range of phenomena, but always by means of more closely specified models.

Of course, some models can also be somewhat abstract—that is, omit certain detailed properties of a phenomenon modeled—but never to the same degree as theories. Although a model may not be *very* specific regarding the properties of a phenomenon, it is still more specific than any theory which is applied in that model. This point is well illustrated by the different models of the pendulum. Some models may be heavily constrained by theories, others not very much or not at all. Models will, however, always be constrained by the phenomenon that is their subject because a model is formulated to provide an interpretative description of a phenomenon, and providing such a description always involves a certain amount of empirical constraints. I attend to this issue in chapter 8, but before this, the picture of the relationships between theory, models, and phenomena is completed further in chapter 7.

Notes

1. For an overview, see Craver 2002. Suppes (1961, p. 176) has commented: "My own conviction is that the set-theoretical concept of a model is a useful tool for bringing formal order into the theory of experimental design and analysis of data. The central point for me is the much greater possibility than is ordinarily realized of developing an adequately detailed formal theory of these matters."

2. Note that Hesse's (1953) first article on models already relies on the discussion of a historical example, in this case nineteenth-century models of the aether.

3. Suppes (1961, p. 166) has written: "It is true that many physicists want to think of a model of the orbital theory of the atom as being more than a certain kind of set-theoretical entity. They envisage it as a very concrete physical thing built on the analogy of the solar system. I think it is important to point out that there is no real incompatibility in these two viewpoints. To define formally a model as a set-theoretical entity which is a certain kind of ordered tuple consisting of a set of objects and relations and operations on these objects is not to rule out the physical model of the kind which is appealing to physicists, for the physical model may be simply taken to define the set of objects in the set-theoretical model."

4. The Semantic View of theories is not uniform but comes in different variants put forward by a number of different authors (van Fraassen 1980 and 1987; Suppes 1961 and 1967; Sneed 1971; Giere 1988; and Suppe 1989). For my purposes, however, it is not necessary to consider these differences in detail. My reservations with this approach are sufficiently general not to rely on the detailed distinctions between the positions of the various proponents of the Semantic View.

5. Craver (2002, p. 65) has captured the same idea by saying: "Theories are not identified with any particular representation." Van Fraassen (1987, p. 109) talks about the "mistake . . . to confuse a theory with the formulation of a theory in a particular language."

6. Some may argue that pictures can also tell us something and that a model may consist of a picture. But as I spell out in chapter 8, pictures often require text to explain how to "read" the picture, and moreover, some of what the picture tells us is likely to be expressible in terms of text (which is not to say that the picture may not, under certain circumstances, communicate the content of the picture more efficiently). My notion of "description" is wide enough to include pictures, (some of) the content of which is expressible in terms of propositions.

7. The idea of models as mediators has been around before, however. Adam Morton (1993, p. 663), discussing models of the atmosphere, talks about "mediating models": "They mediate between a *governing theory*, which

I take to be a true but unmanageable description of some underlying processes and the phenomena which they produce but which the theory does not easily yield."

8. One could also think of phenomenological laws as laws that are merely generalizations, but they would be generalizations only with reference to one particular phenomenon—that is, not "very" general.

9. Although she does not make this explicit in 1983, Cartwright (1999b, p. 242) later highlights the proximity of her early position on scientific models to the Semantic View of theories: "*How the Laws of Physics Lie* supposed, as does the semantic view, that the theory itself in its abstract formulation supplies us with models to represent the world." There may be a similarity in how to characterize the relationship between models and theory, but the fact that Cartwright portrays models as descriptions, even if fictional, does not go together with the portrayal of models as nonlinguistic entities that they are according to the Semantic View.

10. Cartwright seems to imply here that position, extension, motion, and mass are concrete concepts, or at least more concrete than force. This seems like a claim hard to defend, but I leave this issue here.

11. For a critique of the treatment of this example, see French and Ladyman 1997.

12. Compare the different stages of modeling a pendulum, from the ideal pendulum to more and more "realistic" versions of pendulums, as discussed in section 6.6.

13. For a detailed discussion of the distinction between representative and interpretative models in the context of the BCS model of superconductivity, see Morrison 2008.

14. Cartwright (1989) discusses idealization and abstractness in chapter 5 of *Nature's Capacities and Their Measurement.* There, the notions of abstractness and idealization are expected to be useful in the context of the concept of capacities and of causality, but this is a somewhat different context from *theories* being abstract.

15. There are exceptions to this. For some phenomena that are modeled, there exists only one specimen that is taken into account—for example, the Earth or our solar system.

16. I am aware that the term "prototype" has some connotations that are counterintuitive to my use of it, but for want of a better alternative, I introduce it here as a technical term to be used in the way described here in the chapter.

17. This definition of abstraction is contrasted in the next section with a definition of idealization.

18. The *Oxford English Dictionary* gives for "abstract": "Withdrawn or separated from matter, from material embodiment, from practice, or from particular examples. Opposed to *concrete*," in addition to older uses. Cartwright (1989, pp. 197 and 213ff.) identifies this as the Aristotelian notion of abstrac-

tion. She recounts: "For Aristotle we begin with a concrete particular complete with all its properties. We then strip away—in our imagination—all that is irrelevant to the concerns of this moment to focus on some single property or set of properties, 'as if they were separate'" (ibid., p. 197). See also Chakravartty 2001, pp. 327ff.

19. For a case study supporting this kind of "division of labour" between models and theories, see Suárez 1999.

20. Incidentally, in one physics textbook the authors go on to discuss how the acceleration due to gravity in the case of the Earth can be measured using a pendulum. They nicely highlight the contrast I am trying to emphasize between modeling a real and modeling a fictional thing by saying that "[a pendulum used for measuring the acceleration due to gravity] must necessarily be a physical pendulum, since the mathematical pendulum is only an idealisation" (Frauenfelder and Huber 1966, p. 177).

21. See also Giere 1988, pp. 76ff. Giere (1994) relates the organizational structure of classical mechanics to findings on concept formation in cognitive science. His proposal is that there is a family of models of different pendulums, although with a graded structure. There are central examples of pendulums and peripheral ones. While the ideal pendulum constitutes the focal model, peripheral examples include physical, damped, driven, and coupled pendulums. For the physical pendulum, the moment of inertia around the pivot point needs to be calculated as well as the length of the equivalent ideal pendulum. For the damped or driven pendulums, additional forces need to be included in the equation of the movements of a pendulum. Moreover, a physical pendulum can also be driven or damped, or both. The peripheral pendulums are those of increased complexity, and interestingly, these are the ones that get closer to real phenomena; one deals with "a complexity introduced in order better to represent actual physical systems" (ibid., p. 287). Correspondingly, the "focal" model of the ideal pendulum is the one farthest away from actual physical systems.

22. Interestingly, Suárez (1999) employs just this example of the pendulum to argue that using idealization, and subsequently de-idealization, in order to apply theory to phenomena makes models appear superfluous in comparison to theories. I tried to illustrate here, in contrast, that modeling a real pendulum involves more than merely de-idealization: fitting bits of different theories together to approach a fairly realistic account of the phenomenon.

References

Bailer-Jones, D. M. 1999. Creative strategies employed in modelling: A case study. *Foundations of Science* 4: 375–388.

———. 2000. Modelling extended extragalactic radio sources. *Studies in History and Philosophy of Modern Physics* 31B: 49–74.

———. 2002. Sketches as mental reifications of theoretical scientific treatment. In M. Anderson, B. Meyer, and P. Olivier, eds., *Diagrammatic Representation and Reasoning.* London: Springer-Verlag, pp. 65–83.

Bartels, A. 2002. Fiktion und Repräsentation (Fiction and representation). *Zeitschrift für Semiotik* 24: 381–395.

Beatty, J. 1980. Optimal-design models and the strategy of model building in evolutionary biology. *Philosophy of Science* 47: 532–561.

———. 1982. What's wrong with the Received View of evolutionary biology? In P. D. Asquith and R. N. Giere, eds., *PSA 1980, Proceedings of the 1980 Biennial Meetings of the Philosophy of Science Association.* Volume 2. East Lansing, Michigan: Philosophy of Science Association, pp. 397–426.

Beth, E. 1949. Toward an up-to-date philosophy of natural sciences. *Methodos* 1: 178–185.

Boumans, M. 1999. Built-in justification. In M. Morgan and M. Morrison, eds., *Models as Mediators.* Cambridge: Cambridge University Press, pp. 66–96.

Burian, R. M. 2001. The dilemma of case studies resolved: The virtues of using case studies in the history and philosophy of science. *Perspectives on Science* 9: 383–404.

Cartwright, N. 1983. *How the Laws of Physics Lie.* Oxford: Clarendon Press.

———. 1989. *Nature's Capacities and Their Measurement.* Oxford: Clarendon Press.

———. 1991. Fables and models. *Proceedings of the Aristotelian Society* (suppl.) 65: 55–68.

———. 1997. Models: The blueprints for laws. In L. Darden, ed., *PSA 1996. Philosophy of Science* 64 (Proceedings): S292–S303.

———. 1998. How theories relate: Takeovers or partnerships? *Philosophia Naturalis* 35: 23–34.

———. 1999a. *The Dappled World: A Study of the Boundaries of Science.* Cambridge: Cambridge University Press.

———. 1999b. Models and the limits of theory: Quantum Hamiltonians and the BCS model of superconductivity. In M. Morgan and M. Morrison, eds., *Models as Mediators.* Cambridge: Cambridge University Press, pp. 241–281.

———, T. Shomar, and M. Suárez. 1995. The tool box of science: Tools for the building of models with a superconductivity example. In W. E. Herfel, W. Krajewski, I. Niiniluoto, and R. Wojcicki, eds., *Theories and Models in Scientific Processes.* Poznan Studies in the Philosophy of the Sciences and the Humanities. Amsterdam: Rodopi, pp. 137–149.

Chakravartty, A. 2001. The semantic or model-theoretic view of theories and scientific realism. *Synthese: An International Journal for Epistemology, Methodology, and Philosophy of Science* 127: 325–345.

Craver, C. 2002. Structures of theories. In P. Machamer and M. Silberstein, eds., *The Blackwell Guide to the Philosophy of Science.* Oxford: Blackwell, pp. 55–79.

Frauenfelder, P., and P. Huber. 1966. *Introduction to Physics.* Vol. 1. Oxford: Pergamon Press.

French, S., and J. Ladyman. 1997. Superconductivity and structures: Revisiting the London account. *Studies in History and Philosophy of Modern Physics* 28: 363–393.

Giere, R. 1988. *Explaining Science: A Cognitive Approach.* Chicago: University of Chicago Press.

———. 1994. The cognitive structure of scientific theories. *Philosophy of Science* 61: 276–296.

———. 1999. Using models to represent reality. In L. Magnani, N. Nersessian, and P. Thagard, eds., *Model-Based Reasoning in Scientific Discovery.* New York: Plenum Publishers, pp. 41–57.

Hesse, M. 1953. Models in physics. *British Journal for the Philosophy of Science* 4: 198–214.

McKinsey, J. C. C., and P. Suppes. 1953a. Axiomatic foundations of classical particle mechanics. *Journal of Rational Mechanics and Analysis* 2: 253–272.

———. 1953b. Transformations of systems of classical particle mechanics. *Journal of Rational Mechanics and Analysis* 2: 273–289.

McMullin, E. 1985. Galilean idealization. *Studies in History and Philosophy of Science* 16: 247–273.

Morgan, M., and M. Morrison, eds. 1999. *Models as Mediators.* Cambridge: Cambridge University Press.

Morrison, M. 2008. Models as representational structures. In S. Hartmann, C. Hoefer, and L. Bovens, eds., *Nancy Cartwright's Philosophy of Science.* New York: Routledge.

Morrison, M. C. 1998. Modelling nature: Between physics and the physical world. *Philosophia Naturalis* 35: 65–85.

———. 1999. Models as autonomous agents. In Morgan and Morrison, *Models as Mediators.* Cambridge: Cambridge University Press, pp. 38–65.

———, and M. S. Morgan. 1999. Models as mediating instruments. In Morgan and Morrison, *Models as Mediators.* Cambridge: Cambridge University Press, pp. 10–37.

Morton, A. 1993. Mathematical models: Questions of trustworthiness. *British Journal for the Philosophy of Science* 44: 659–674.

Pinnick, C., and G. Gale. 2000. Philosophy of science and history of science: A troubling interaction. *Journal for General Philosophy of Science/Zeitschrift für allgemeine Wissenschaftstheorie* 31: 109–125.

Pitt, J. 2001. The dilemma of case studies: Toward a Heraclitian philosophy of science. *Perspectives on Science* 9: 373–382.

Rouse, J. 2003. Kuhn's philosophy of scientific practice. In T. Nickles, ed., *Thomas Kuhn.* Cambridge: Cambridge University Press, pp. 101–121.

Sneed, J. 1971. *The Logical Structure of Mathematical Physics.* Dordrecht: Reidel.

Suárez, M. 1999. The role of models in the application of scientific theories: Epistemological implications. In M. Morgan and M. Morrison, eds., *Models as Mediators.* Cambridge: Cambridge University Press, pp. 168–196.

Suppe, F. 1974. The search for philosophic understanding of scientific theories. In F. Suppe, ed., *The Structure of Scientific Theories.* Urbana: University of Illinois Press, pp. 3–241.

———. 1989. *The Semantic Conception of Theories and Scientific Realism.* Urbana: University of Illinois Press.

———. 2000. Understanding scientific theories: An assessment of developments. In D. Howard, ed., *PSA 1998, Philosophy of Science Supplement* 67, pp. S102–S115.

Suppes, P. 1961. A comparison of the meaning and uses of models in mathematics and the empirical sciences. In H. Freudenthal, ed., *Concept and the Role of the Models in Mathematics and Natural and Social Sciences.* Dordrecht: D. Reidel Publishing Company, pp. 163–177.

———. 1962. Models of data. In E. Nagel, P. Suppes, and A. Tarski, eds., *Logic, Methodology, and Philosophy of Science.* Stanford, California: Stanford University Press, pp. 252–261

———. 1967. What is a scientific theory? In S. Morgenbesser, ed., *Philosophy of Science Today.* New York: Basic Books Publishers, pp. 55–67.

van Fraassen, B. 1980. *The Scientific Image.* Oxford: Clarendon Press.

———. 1987. The semantic approach to scientific theories. In N. J. Nersessian, ed., *The Process of Science.* Dordrecht: Martinus Nijhoff Publishers, pp. 105–124.

———. 1991. *Quantum Mechanics: An Empiricist's View.* New York: Oxford University Press.

Wimsatt, W. 1987. False models as means to truer theories. In M. Nitecki, ed., *Neutral Models in Biology.* Oxford: Oxford University Press, pp. 23–55.

PHENOMENA, DATA, AND DATA MODELS 7

I DEFINED A MODEL as an interpretative description of a phenomenon in chapter 1, section 1.1. I said that "phenomenon" refers to events, facts, and processes and tacitly relies on an intuitive understanding of what a phenomenon is, and did so without providing further discussion. In this chapter, I rectify this omission by exploring what phenomena are and how they are constituted. Moreover, I examine the relationship between a phenomenon and the data derived from that phenomenon. Models, being about phenomena, are expected to match the available empirical data derived from the phenomena. The relationship between data and phenomena is, however, not always straightforward. In this chapter, I outline the relationships between phenomena, data, data models, and theory to illustrate the role models play regarding phenomena.

Looking closer at how experiments are conducted and what kind of data they produce, as the so-called new experimentalists did (for example, Hacking 1983; Galison 1987; and Franklin 1989), shifted the focus of scientific methodology to a consideration of how scientific data are acquired. This led to the fairly recent distinction between data and phenomena by James Bogen and James Woodward, which I introduce and develop in section 7.1. This discussion of data shows that there is no simple notion of experimentally confirming a theory.

In contrast, as reviewed in section 7.2, there are many steps of processing data and even making data suitable for comparison with theoretical results. How a phenomenon may "change" in the course of being modeled is discussed in section 7.3. Some authors propose that there is a whole hierarchy of models (section 7.4). Correspondingly, scientific models need to be repositioned in the whole of the scientific process and in scientific methodology. The resulting "complete" picture is presented in section 7.5, a picture that retraces the links from phenomenon to data about the phenomenon, from data to a scientific model of the phenomenon and from the scientific model to the phenomenon. My main claim is that what we take a phenomenon to be is significantly influenced by the way we model it.

7.1 What Is a Phenomenon?

In my understanding of phenomenon, I largely follow the work of Bogen and Woodward (1988, 1992, and 2003; see also Woodward 1989 and 2000). A phenomenon is a fact or event in nature, such as bees dancing, rain falling, or stars radiating light. A phenomenon is not necessarily something as it is observed; Bogen and Woodward's point is precisely to distinguish data about a phenomenon from the phenomenon. A phenomenon may be something that is originally picked up by observation and then raises certain questions. Observing a bee's dance—and even calling it that—may bring about the conjecture that there is something systematic about the bee's movements that warrants further investigation. To conjecture thus is not to take the bee's movements as something happening entirely at random. It is *treating* what is observed as a phenomenon. So, in the very first instance, a phenomenon is something that is taken to be a subject to be researched. At this stage, it is not strictly known whether there really is a distinguishable fact or event to be found, even if one has an inkling that this is so. Similarly, if we observe rain falling, asking what causes the rain is the first step of turning this observation into something that constitutes a phenomenon. This seems to suggest that picking out a phenomenon has something to do with distinguishing the causal processes that make up that phenomenon. This understanding is, I think, in accordance with the understanding of "phenomenon" that Ian Hacking (1983, p. 221) has attributed to scientists: "[Phenomenon] has a fairly definite sense in the writing of scientists. A phenomenon is *noteworthy*. A phenomenon is *discernible*. A phenomenon is commonly an event or process of a certain type that occurs regularly under definite circumstances."

At an early stage, a phenomenon is something about which one wants to know more. Then, in the process of learning and discovering more about a phenomenon, what the phenomenon is taken to be and to consist of changes. Take gold, for example. Gold is a material that was originally identified probably by its color and some of its properties. In a larger theoretical context it is the element on the periodic table that has the atomic number 79. Gold can be involved in various physical processes constituting phenomena, such as chemical reactions, and it is in the context of these reactions that the atomic number of gold receives its significance. It turns out that this atomic number is inseparably linked to how gold is involved in natural processes that constitute certain phenomena. This means, for instance, that gold behaves and reacts in a way that is comparable to other elements of the same main group of the periodic table, such as copper and silver. One phenomenon is constituted by the fact that metals of this group do not corrode as easily as other metals, such as iron. So the study of gold that places gold in a certain theoretical context changes how one would delineate phenomena involving gold.

Certain other phenomena (not involving gold) would not even be recognized without at least some basic research into them. Examples for such phenomena are nuclear fusion or the order of acquisition of prepositions by children. It takes considerable observation and experimentation to establish that children acquire the meaning of *in* before *on* and finally *under*. Once this is established as a phenomenon in some languages (for example, English and German), it becomes possible to examine whether this finding is more universal and holds for other, structurally different languages (such as Polish) too (Rohlfing 2002). If it were found that this order of acquisition of prepositions held for all languages—that is, appeared to be cognitively universal—then this finding would change somewhat what the phenomenon is. The phenomenon would no longer be something that occurs in selected languages. In fact, for a phenomenon like this, without prior examination it is not even obvious that anyone could recognize it as a phenomenon.

Sensual perception is certainly not in all instances enough to identify and establish a phenomenon. Correspondingly, Bogen and Woodward (1988, p. 352) have acknowledged: "It is overly optimistic, and biologically unrealistic, to think that our senses and instruments are so finely attuned to nature that they must be capable of registering in a relatively transparent and noiseless way all phenomena of scientific interest, without any further need for complex techniques of experimental design and data analysis." Thus that something is identified as a phenomenon is, in many instances, already the result of research. Of course, even if the data are to be distinguished from phenomena, as Bogen and Woodward argue, one expects that a phenomenon manifests itself empirically somehow—that phenomena can become noticed in the empirical world. This brings up the issue of the data associated with a phenomenon, although Bogen and Woodward (ibid.) are adamant that these data must not be identified with the phenomenon itself.

Why this careful phrasing of data being associated with phenomena? Let me illustrate the problem with an example. Bogen and Woodward (ibid.) adopted this example from Ernest Nagel's ([1960] 1979, p. 79) *Structure of Science.* The topic is the melting point of lead. This can be measured and found to be 327 degrees Celsius. What does it mean, however, to find the melting point to be this precise temperature? To establish the melting point of lead, data are collected: a whole series of measurements are carried out. The temperature of melting is typically not measured only once, but many times to take account of measuring errors that are expected. It can happen in principle that, during the whole series of measurements, the precise value of 327 centigrade is never once read off the thermometer. Instead, the average of the measured values is taken, the measurement error calculated, and the result of this data analysis declared the melting point of lead. Because measuring errors occur in exploring nature, measurements have to be repeated many times. As a consequence, large amounts of data are produced that need to be interpreted and analyzed in a way that allows scientists to extract a definite empirical

Element			Melting Temperature
C			3550° C
Si	metallic ↓	↑ covalent	1414° C
Ge	binding ↓	↑ binding	938.25° C
Sn	increases ↓	↑ decreases	231.93° C
Pb			327.46° C

FIGURE 7.1. Binding energy and its relation to temperature in some elements.

finding about a phenomenon (that is, one value for the melting point of lead), albeit with a finite, calculable uncertainty.

Let me now look at the phenomenon of lead melting at a certain temperature, which, as it is claimed, differs significantly from data about the melting point. This phenomenon is also about the factors that make up that melting point. One can, for instance, ask why the melting point is as high or as low as it is. At a theoretical level, this question has to do with the forces that hold together the atoms and molecules and so influence the melting point.[1] There are different forces and correspondingly different models of chemical binding. For different chemical elements, different forces are pronounced—for example, the London forces in crystals, hydrogen bridge binding in water, or ion binding in metals. The melting point of a specific element depends on the binding forces that act in the case of that element. One has to ask which binding force is most central in determining the melting point. For this, the comparison with similar elements, those of the same main group of the periodic table, is relevant. In the case of lead, this is the fourth main group of the periodic table: carbon, silicon, germanium, tin, and lead. In this group, covalent binding decreases and metallic binding increases. Correspondingly, lead is subject to stronger metallic binding than tin, which is why it has a higher melting point, whereas both have much lower melting points than carbon, silicon, and germanium because these latter elements have much stronger covalent binding (figure 7.1).

Although these theoretical considerations may not suffice to calculate the melting point of lead, they nonetheless allow us to appreciate what constitutes the phenomenon of the melting point of lead.

Part of this is to be able to explain and predict a certain behavior of lead, something that is not possible purely on the basis of measuring the melting point. Such systematic explanation of facts about phenomena is, according to Bogen and Woodward, precisely what can be achieved with theories. The phenomenon of the melting point of lead is thus conceptually different from data about lead that are produced when one tries to establish the melting point of lead experimentally. Different ramifications apply to the data analysis and the theoretical description of the phenomenon.

7.2 Data

Bogen and Woodward (1988) explain the discrepancy between data and phenomena by pointing out that a large number of causal factors play a role in experiments, factors that depend on the specific experimental setup, on the kind of measuring apparatus, and the methods of data processing and data analysis. In turn, as phenomena, only a relatively small and manageable number of causal factors are considered. They write: "In undertaking to explain phenomena rather than data, a scientist can avoid having to tell an enormous number of independent, highly local, and idiosyncratic causal stories involving the (often inaccessible and intractable) details of specific experimental and observational contexts. He can focus instead on what is constant and stable across different contexts" (ibid., p. 326).

While data depend on the specific experimental situation that needs to be checked for measuring errors, the phenomenon that is investigated has to display stable characteristics that can be established repeatedly and by means of different experiments. There can be vastly different experimental procedures that each lead to the confirmation of the melting point of lead, and depending on the experimental procedures, data need to be interpreted quite differently. Bogen and Woodward (ibid., p. 327) point out that it is possible to track down a measuring error and its size without knowing the source for that error. No matter how complicated and intricate the experimental confirmation, the phenomenon is independent of the individual data processing procedure and measuring procedure, even though the data derive from the phenomenon via an experimental procedure. Such stability arises because the number of factors that are taken to constitute the phenomenon is limited.

Bogen and Woodward (1992) turn against the suggestion that data might be theory-laden because of the involvement and the bias of human observers. Their point is that the reliability of data can be tested quite irrespective of how measurements are made and how much human "input" they involve. Needless to say, human observers are involved in the data production process, but human observation does not make up the whole of the data production process. Humans may be needed for setting up the measuring apparatus, reading values off a dial, and so on, but it would be the exception if most of the measuring were done by the human observer. Of course, readings from meters can be biased, but there are often ways of reducing and controlling that bias. There are statistical checks and practical tricks, such as putting a mirror behind the pointer needle to avoid reading the dial at an angle or using a number of independent observers. Correspondingly, human perceptual shortcomings are unlikely to be a major cause of unreliable data.[2] Usually, the human observer constitutes but a small link somewhere in the data production process. Bogen and Woodward (ibid., p. 597) write: "[A]lthough perception (and worries about the reliability of perception, and the control of perceptual error) are sometimes central to data-interpretation, the reliability of the procedures used to move from data to phenomena typically has more to do with nonperceptual (e.g., statistical) techniques of data analysis, and with strategies for the control of nonperceptual error."

7.3 How to Identify Phenomena

Earlier in this chapter, I introduced a phenomenon as a subject of research. I since indicated that the knowledge of what constitutes a certain phenomenon is also a result of research previously carried out, involving theorizing and identifying causal factors. This means that the establishment of a phenomenon has links both to theory and to data. This is the point at which James McAllister (1997) criticizes

Bogen and Woodward.[3] The latter assume that phenomena are natural kinds of which the world consists. McAllister takes them to assume that phenomena can be discovered as patterns in data sets.[4] He objects to this proposal because for each data set there exist infinite possibilities to interpret that data set as "pattern A + noise at *m* percent"—that is, as any pattern with a certain error. Therefore, criteria are required that permit one to identify those patterns (plus accepted error) that constitute phenomena.

According to McAllister, such criteria are chosen and applied by the researchers who analyze the data set. If phenomena were natural kinds, as Bogen and Woodward claim, then, *pace* McAllister, phenomena could not be entities the establishment of which is subject to such choices as involved in identifying a pattern and identifying the acceptable noise level. This is why McAllister views Bogen and Woodward's account as incoherent: on the one hand, there exist infinitely many patterns, and on the other hand, a phenomenon is supposed to stand out as one particular pattern in a data set. Already existing theories cannot aid the selection procedure because theories should only come in to describe and explain phenomena after the phenomena are identified independently in the data set. This is why theories cannot be responsible for the definition or identification of phenomena from data, if a vicious circle is to be avoided.[5] McAllister concludes from this that the identification of phenomena is based on a decision taken by the researcher and cannot be independent of such a decision, as Bogen and Woodward suggest. Correspondingly, what counts as a phenomenon depends on the investigator of that phenomenon.

McAllister claims for his position the advantage that it does justice to historical change of what counts as a certain phenomenon. His example is that Kepler and Newton each saw different patterns and errors in the data sets, depending on whether they interpreted the patterns as ellipses or as planetary orbits due to gravitation. McAllister (1997, p. 226) writes: "Which aspect of the occurrences is singled out for explanation varies from one investigator to another, notwithstanding the fixedness of physical occurrences themselves." This would imply that phenomena are theory-laden, just as the example of the phenomena associated with gold suggests. Of course, phenomena manifest themselves in the form of data in one way or another, but the application of theories to these phenomena plays a role in the choice of the relevant patterns in the data and thus in establishing a phenomenon.

Perhaps it is misconceived to think of phenomena as effectively patterns in data sets. My alternative suggestion is to identify a phenomenon with recognizing that something has the potential to be theoretically explained. This can, in principle, be the case before any data are collected. It is for the first time to view the bee's bustle as something with a system to it, something that has a significance, that can be examined and worked out. If data has not been collected, it is also not possible to identify the phenomenon (at that stage anyway) on the basis of a pattern in the data set. Even later, if a phenomenon is more fully interpreted, what are taken to be the facts about something may change with one's theoretical interpretation of what one is observing. This sounds somewhat similar to Hacking (1983, pp. 220ff.) talking of the "creation of phenomena," but in his case phenomena are *created* because they only occur under very specific, human-made experimental conditions. Hacking's example is the Hall effect, of which he says that it does not exist outside a certain apparatus: "I suggest, Hall's effect did not exist until, with great ingenuity, he had discovered how to isolate, purify it, create it in the laboratory" (ibid., p. 226).

P. Kroes (1995, p. 435) considers a weak and a strong interpretation of Hacking's claim. The weak interpretation has to do with creating the proper experimental conditions for a phenomenon to occur. The strong interpretation is an ontological claim and relates to creating the very features of the phenomenon itself. Kroes argues convincingly in his article that Hacking can only have the weak interpretation in mind. Similarly, I do not suggest that phenomena only begin to exist, in an ontological sense, when they attract our interest. This is, however, when we start to consider them as phenomena. Moreover, in contrast to Hacking, the kinds of phenomena I consider may just as well occur in a natural environment as in an artificially prepared experimental environment.

Not only is it not always clear how to delineate a phenomenon, shedding doubt on the claim that phenomena are natural kinds, but it is also sometimes hardly possible to interpret data without having a phenomenon in mind. Prajit Basu (2003, p. 356) presents a nice case study illustrating just this point: "observations, when transformed into evidence for a hypothesis, phenomena, or a theory, are theory infected." Basu considers a case where two researchers perfectly agree on the data, but they take it to be evidence for different phenomena. The two researchers are Antoine-Laurent Lavoisier (1743–1794) and Joseph Priestley (1733–1804). Lavoisier wanted to argue that water is a compound. One indication was that hydrogen and oxygen react together to form water. The other indication was that iron and steam (water) react to produce iron oxide and hydrogen (that is, in this reaction, water is split up into oxygen and hydrogen).[6]

While Priestley had no doubts concerning the observational side of this reaction, he doubted that the black powder which Lavoisier took to be iron oxide was in fact iron oxide (see also McEvoy 1988, p. 208). The lesson Basu (2003, p. 357) draws from this example is the following: "A piece of evidence for (or against) a theory is a construction in the context of that theory from (raw) data. In this construction, a set of auxiliary assumptions is employed. These auxiliaries may themselves be theoretical in character. From the same (raw) data it is possible to construct different evidence for (or against) different theories since the auxiliaries employed in connection with different theories can be different. Finally, although the (raw) data are expressed in a language which is acceptable to partisans of competing theories, the evidence constructed from the same (raw) data is often expressed in the partisans' differing theoretical languages."

The fact that the chemical reaction I described earlier resulted in a black powder (raw data) both Priestley and Lavoisier could agree on, but they could not agree on what that black powder could be taken as evidence for. Priestley claimed that in addition to the black powder, a gas was formed during the reaction. Lavoisier accepted the principle of the conservation of mass, that the weights of what went into the reaction were the same as the weights of what came out. On the basis of this (theoretical) principle, Lavoisier had reason to argue that there was no gas in addition to the black powder. (This is evidence rather than data.) Basu even identifies a number of levels of evidence. For instance, even to establish that the sample of iron is pure requires a test that is based on certain theoretical assumptions. At this lower level, Priestley agreed with Lavoisier, but obviously evidence could be questioned at any level. Agreement may be required even to establish to which level something counts as raw data. Basu (ibid., p. 364) then concludes: "To the extent that these (raw) data are transformed into evidence, and for any evidential bearing these data might have on a particular theory and hence any bearing they might have on theory resolution, the evidence is theory-laden." So in a way, depending on their theoretical assumptions, Priestley and Lavoisier could have taken the same data, on which they agreed (namely, black powder), as evidence for different phenomena.[7] There may not be any problem with raw data, but using raw data as evidence for a phenomenon is difficult without having the particular phenomenon in mind.

With my earlier examples concerning gold, lead, and the order in which prepositions are acquired, I tried to emphasize the point that Basu has worked out so vividly with his example from chemistry. Phenomena are likely to depend on theoretical assumptions when they are established. What counts as a phenomenon may also mutate from what originally gave rise to its examination to something that increasingly incorporates already existing theoretical and empirical results. This understanding of a phenomenon is then expressed in terms of one or several models that provide a description of the phenomenon. To provide such a description, models rely, on the one hand, on known empirical facts derived from the data and, on the other hand, on theories that are applied in the model.

Accordingly, Margaret Morrison (1999, for example, p. 45) has claimed that models are able to mediate between theory and the world and intervene in both domains. Correspondingly, models and phenomena are closely tied together, whereby the phenomenon is centered on what is found in nature and the model on how these findings are captured and described with reference to accepted theories. The link between the two becomes tighter the better researched a phenomenon is. So the model may provoke experiments for the further examination of the phenomenon. In turn, when at the outset there exists no model for a phenomenon and instead only some data about the phenomenon, then it remains up to the discretion of the researcher with his or her assumptions and theoretical hypotheses to take seriously certain patterns in the data and to try and design a model on their basis. In the course of these modeling attempts and modeling successes, what constitutes the phenomenon is refined and

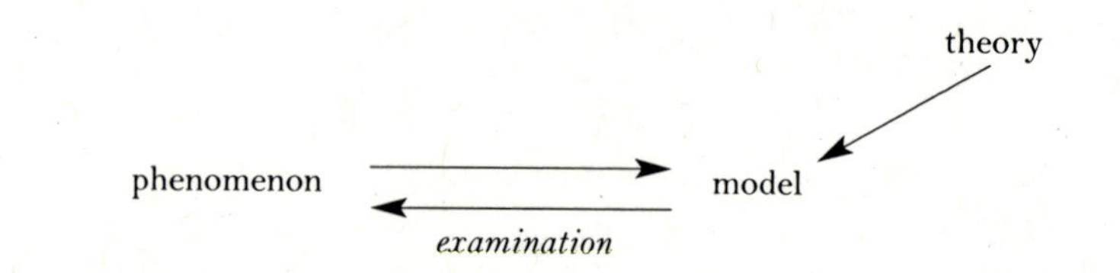

FIGURE 7.2. In order to capture a phenomenon, the phenomenon is modeled, and the way the phenomenon is modeled will influence how the phenomenon is defined. Theory provides the background for the model development.

redefined more and more. A simple illustration of this process is found in figure 7.2. .

What can Bogen and Woodward say in defense of phenomena as natural kinds? First, the claim is that phenomena exist independently of being investigated. They are out in the world, whether discovered or not. They are effects that can (and in many cases do) occur repeatedly as a result of the same (relatively small) causal influences. As Bogen puts it, Jacksonian epileptic seizures occurred long before Jackson began to study them and were not changed by his investigation of them (Bogen, personal communication with the author). However, I would reply that such phenomena can often only be *identified* when certain theoretical considerations enter the debate and when certain causal factors constituting these phenomena are already identified. Yes, they may exist "untouched," but they are often not identified and known about without being touched. In my books, being a phenomenon is not possible without being recognized as one.

7.4 Hierarchies of Models

The distinction between data and phenomena also shows that it is a long way from data (that are supposed to confirm and establish a phenomenon) to the phenomenon itself, simply because data alone do not constitute a phenomenon. As the example in the previous section illustrates, this is especially evident when dealing with the statistical analysis of data. Considering the statistical analysis of data originally led Patrick Suppes (1962) to the introduction of the concept of a *data model* and to assuming a whole hierarchy of models. He wrote: "[E]xact analysis of the relation between empirical theories and relevant data calls for a hierarchy of models of different logical type" (ibid., p. 253).

Suppes was concerned with a formal analysis of group-theoretical relationships between different types of models. Deborah Mayo then adopted his idea of a hierarchy of models (and so did others, for example, Giere 1999 and Harris 1999), but this time in the context of thinking about experiments (Mayo 1996, p. 131). The point is to make the different steps of analysis explicit that have to take place in order to connect "raw" data, produced by an experiment, with a scientific hypothesis that is expressed by a theoretical model ("primary theoretical model of the inquiry" [ibid., p. 133]). Roughly speaking, the hierarchy of models consists of data model, model of the experiment, and theoretical model.

A *model of the experiment* forms the link between the model of data and the theoretical model. This model is about how to test experimentally a hypothesis stated in a theoretical model. In this capacity the model of the experiment still omits many practical problems that may arise in the experiment. For instance, the model may employ idealizations, such as frictionless planes. The model is about the conception of the experiment and not about the actual empirical results that the experiment may produce. So, examining the melting point of lead, the model of the experiment would be about the method with which to measure the temperature of lead to produce a useful, reliable result.[8]

A *data model* has the task to put raw data into a canonical form, which makes it possible to compare the experimentally produced data with the prediction of the theoretical model (for example, how to analyze the individual measurement values of the temperature of lead statistically). The point is that data can only be used after the application of a data analysis technique. The data model still has nothing to do with any theoretical assumptions that may go into modeling phenomena. An example for the different steps of interpretation is observing a celestial object with a radio telescope. To achieve high resolution, large diameter telescopes are necessary. There are, however, practical limits to the size of a telescope dish, which is why the technique of aperture synthesis is employed. Instead of using one large telescope dish, one measures the phase differences between radio waves, in addition to the intensity of the waves, received by a number of telescopes, with smaller dishes, positioned at a distance from each other. Using interferometry, this phase information makes it possible to reconstruct the image that would have been received if a single telescope of much larger diameter had been used. So the conceptual idea of aperture synthesis would have to be captured in a model of the experiment (or, strictly, of the observation instrument).

Data analysis means in this case that the data from the different telescopes involved must be synthesized—that is, brought together, corrected for measuring errors, and interpreted such that they become equivalent to the observational result that could have been achieved with only one telescope with an extremely large dish. How the data from the different telescopes are to be processed is captured in data models. Often the result of such data analyses is presented in the form of a radio contour map, where the lines on the map connect areas of the same intensity of radiation and lines close together indicate where the intensity is changing rapidly. Such maps that display the intensity distribution of an object can then be used to test hypotheses about a phenomenon that are expressed in a theoretical model of the phenomenon—for example, to examine whether double radio sources have one or two jets or whether the lobes of a radio source contain older plasma than the hotspots, and so on (cf. Bailer-Jones 2000). The hierarchy of models laid out in this section plays its role in the larger picture of the modeling process that is presented below.

7.5 The "Complete" Picture

The considerations presented in this chapter allow me to sketch a picture of how data, phenomena, and theory are connected via various types of models. Let me summarize the claims thus far. Data result from the examination of phenomena, of objects of empirical research. Examining a phenomenon can mean observing the phenomenon or experimenting on it (whereby the outcome of the experiment also has to be observed one way or another). Observation requires skill, and such observational skill need not be directly linked to the theoretical considerations that led to the construction of the experiment—the model of the experiment—or to the construction of observational apparatus, such as telescopes, microscopes, video cameras, Geiger-Müller counters, and so on. Once data are produced, one needs to confront the issue that the data are riddled with measurement errors and deviations that need to be corrected. The result of this process of data analysis is a data model. Only once the raw data is "translated" into a data model is it possible to test theoretical hypotheses about the phenomenon that provided the point of departure for the data collection. Furthermore, I doubt that it is possible to define phenomena without any reference to theory because the delineation and definition frequently seem to rely on already established research results about the phenomenon, and these are formulated at least partly with reference to a theoretical background.

The resulting picture is as follows: A phenomenon is experimentally or observationally examined. In the first instance, a phenomenon may be an object encountered in nature or the human environment. Yet how to capture this phenomenon also increasingly depends on its empirical examination and theoretical description to this point—that is, on one or more existing theoretical models about the phenomenon. The theoretical model is an attempt to capture the phenomenon by providing an as complete as possible description of it, which includes highlighting those factors that are relevant for constituting the phenomenon. Data about a phenomenon can be produced by a whole range of different experimental procedures. They therefore present individual or isolated evidence about the phenomenon, while the phenomenon itself is expected to display a certain robustness in the face of the different experimental situations. Raw data cannot serve to confirm a theoretical model about a phenomenon, but have to undergo procedures of data analysis and be put into the form of a data model to be usable for an empirical test. Thus empirical confirmation takes place between data model and theoretical model, not between data and phenomenon, and also not between data and theoretical model. This means that the links between data and data model and theoretical model and phenomenon need to be considered in addition.

Figure 7.3 shows that between data and theory there lie a number of steps. The theoretical model plays an important role in applying theory to phenomena because it provides the link to the data model of data derived from the phenomenon. The theory is applied to the phenomenon only via the theoretical model, and the theory is only ever confirmed via the theoretical model that links it to the data model. So theories are linked to empirical findings indirectly: via the

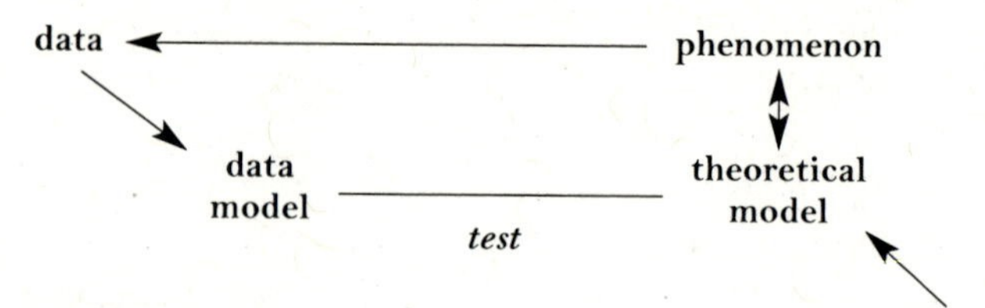

FIGURE 7.3. As in figure 7.2, phenomenon and theoretical model remain closely connected, but the test of the model for a phenomenon takes a "detour" via data generation and data modeling. When the phenomenon is examined experimentally or observationally, data about the phenomenon are produced. To compare the data with the theoretical models, a data model is required. To extract the data from the phenomenon, a model of the experiment or of the observational apparatus needs to be inserted between data and phenomenon (omitted in this diagram).

theoretical model *and* via the data model (and whatever else lies in between). Finally, the way a phenomenon is delineated is closely tied to the theoretical model about the phenomenon, which in turn is shaped by theoretical assumptions that go into its formulation.

A scientific model and its phenomenon are closely connected from the start in that the model is designed as a model *of* the phenomenon. What is taken to be the phenomenon becomes somewhat reconstructed in the course of the modeling process. The modeled phenomenon may depart somewhat from what the phenomenon started out to be taken to be when it first attracted curiosity and was first studied. Despite this perhaps uncomfortable air of constructivism, however, it is the investigation of the *phenomenon* that results in data about the phenomenon, and the data subsequently serve as a constraint for the model. The link of the model to empirical evidence is required to remain strong. While there is an empirical link, how we delineate and describe a phenomenon is invariably also linked to the way we have learned to model it. What we take a phenomenon to be and how we model it develop together over time.

Notes

1. I am grateful to Rüdiger Stumpf for discussions concerning this example.

2. However, our incomplete understanding of the measurement process can have some impact on what is inferred from the data.

3. Glymour (2000) also has criticized Bogen and Woodward's distinction between phenomena and data, although he thinks that the concept of a phenomenon is unnecessary because causal relations could be hypothesized directly from statistical correlations in a data sample suitably analyzed.

4. Bogen (personal communication with the author) questions the inference from the claim that phenomena are natural kinds to the claim that they must be recognizable as patterns in a data set. The question then remains, however, how phenomena are manifested in the world as natural kinds. The claim that phenomena are natural kinds fails to address how phenomena are discovered and established in the first place.

5. This leaves out the option that the theories involved in identifying a phenomenon may be independent of the theories about the phenomenon.

6. For an account of these experiments, see Carrier 2004.

7. The aim here is specifically not to express a preference for Lavoisier's chemical interpretation over Priestley's. At the time, such a preference would not have been empirically warranted. See, for instance, McEvoy 1988, which emphasizes the overlap between Lavoisier's and Priestley's methodological and ontological practices. See also Carrier 1992.

8. Indeed, Bogen and Woodward (1988, pp. 309–310) describe a number of considerations that have to go into planning the experimental side of the measurement of the temperature of lead.

References

Bailer-Jones, D. M. 2000. Modelling extended extragalactic radio sources. *Studies in History and Philosophy of Modern Physics* 31B: 49–74.

Basu, P. K. 2003. Theory-ladenness of evidence: A case study from history of chemistry. *Studies in History and Philosophy of Science* 34: 351–368.

Bogen, J., and J. Woodward. 1988. Saving the phenomena. *The Philosophical Review* 97: 303–352.

———. 1992. Observations, theories, and the evolution of the human spirit. *Philosophy of Science* 59: 590–611.

———. 2003. Evading the IRS. *Poznan Studies in the Philosophy of the Sciences and the Humanities* 20: 223–256. Available online at http://www.ingentaconnect.com/content/rodopi/pozs.

Carrier, M. 1992. Cavendishs Version der Phlogistonchemie oder: Über den empirischen Erfolg unzutreffender theoretischer Ansätze (Cavendish's version of phlogiston chemistry, or: On the empirical success and inappropriate theoretical approaches). In J. Mittelstraß and G. Stock, eds., *Chemie und Geisteswissenschaften. Versuch einer Annäherung* (Chemistry and social sciences: An attempt at harmonization). Berlin: Akademie Verlag, pp. 35–52.

———. 2004. Antoine L. Lavoisier und die Chemische Revolution. In P. Leich, ed., *Leitfossilien naturwissenschaftlichen Denkens.* Würzburg: Königshausen & Neumann.

Franklin, A. 1989. *The Neglect of Experiment.* Cambridge: Cambridge University Press.

Galison, P. 1987. *How Experiments End.* Chicago: University of Chicago Press.

Giere, R. 1999. Using models to represent reality. In L. Magnani, N. Nersessian, and P. Thagard, eds., *Model-Based Reasoning in Scientific Discovery.* New York: Plenum Publishers, pp. 41–57.

Glymour, B. 2000. Data and phenomena: A distinction reconsidered. *Erkenntnis* 52: 29–37.

Hacking, I. 1983. *Representing and Intervening.* Cambridge: Cambridge University Press.

Harris, T. 1999. A hierarchy of models and electron microscopy. In L. Magnani, N. Nersessian, and P. Thagard, eds., *Model-Based Reasoning in Scientific Discovery.* New York: Plenum Publishers, pp. 139–148.

Kroes, P. 1995. Science, technology, and experiments: The natural versus the artificial. In David Hull, Micky Forbes, and Richard M. Burian, eds. *PSA 1994, Vol. 2.* East Lansing, Michigan: Philosophy of Science Association, pp. 341–440. Available online at https://www.msu.edu/unit/phl/PSA/PS942.htm.

Mayo, D. G. 1996. *Error and the Growth of Experimental Knowledge.* Chicago: University of Chicago Press.

McAllister, J. W. 1997. Phenomena and patterns in data sets. *Erkenntnis* 47: 217–228.

McEvoy, J. 1988. Continuity and discontinuity in the chemical revolution. *Osiris* (2nd series) 4: 195–213.

Morrison, M. C. 1999. Models as autonomous agents. In M. Morgan and M. Morrison, eds., *Models as Mediators.* Cambridge: Cambridge University Press, pp. 38–65.

Nagel, E. [1960] 1979. *The Structure of Science.* Indianapolis, Indiana: Hackett Publishing Company.

Rohlfing, K. J. 2002. UNDERstanding: How infants acquire the meaning of UNDER and other spatial relational terms. Ph.D. dissertation, Bielefeld University. Available online at http://archiv.ub.uni-bielefeld.de/disshabi/2002/0026/_index.htm.

Suppes, P. 1962. Models of data. In E. Nagel, P. Suppes, and A. Tarski, *Logic, Methodology, and Philosophy of Science.* Stanford, California: Stanford University Press, pp. 252–261.

Woodward, J. 1989. Data and phenomena. *Synthese: An International Journal for Epistemology, Methodology, and Philosophy of Science* 79: 393–472.

———. 2000. Data, phenomena, and reliability. *Philosophy of Science Supplement* 67: S163–179.

REPRESENTATION 8

Talk about the issue of representation concerning scientific models is relatively recent in the philosophy of science. Broadly speaking, the topic belongs to the context of the debate about scientific realism. Is the real world the way that science portrays it? This question used to be addressed to scientific theories, but as we have seen, the way theories and models are understood has shifted considerably (see chapter 6). When it is said that a model *represents* something, then this is taken to mean, roughly, that the model tells us what the phenomenon that is the subject of the model is like. That there are considerable difficulties with this approach is the topic of later sections in this chapter.

In section 8.1, I begin with scientists' views on how models relate to reality. Section 8.2 approaches the same topic from the philosophical perspective. One of the major issues is that models often make false claims about the world, another that representation is not necessarily linked to any form of resemblance. In section 8.3, I consider how one could talk of truth and falsity in the context of models and what ramifications this talk has. Under which conditions can we think of models as representing phenomena? Having laid out my own analysis of representation in models, I compare representation of models, in section 8.4, to the analysis of representation in art. Whether representation can be interpreted as an identity of structure of model and phenomenon is examined in section 8.5. Section 8.6 contains the conclusion. The result of my analysis is, rather than a fully fledged account of representation, an account of the constraints that apply to models representing.

8.1 How Models Relate to Reality

First I turn once more to quotations from the interviews on models to document how scientists view the relationship between models and reality (see chapter 1; Bailer-Jones 2002a). There is a distinct understanding among the scientists interviewed that by developing models and testing them it is possible to get closer to what the "real thing" is like. Particle physicist Robert Lambourne clearly expresses such a realism:

> And I come from a rather different perspective, different school, that would take the attitude that physics was actually probing reality and modelling was part of that process. And although I couldn't justify the statement, I would like to think that physicists were closing in on reality and not merely forming a model that reflected part of reality.
>
> *Robert Lambourne, particle physicist*

At the same time, Lambourne acknowledges that there is no way of knowing whether a model really matches reality, despite the intuition that it is possible to come closer and closer to reality:

> Inside many physicists there is a feeling that somehow we are dealing with reality, but there is no principle for establishing that's the truth. There is nothing you can adduce that will show that you are dealing with reality, rather than a model of reality that accords with various experimental points of contact with what's really out there.
>
> *Robert Lambourne, particle physicist*

This is the general point that there can be no proof of realism nor antirealism, merely arguments that might favor one or the other.

Chemist and historian of chemistry Colin Russell qualifies what he means by realism. He makes the point that models may have relationships accurately as they are in reality, yet that it may be possible that a model is accurate in one sense, but not in another—for example, when a three-dimensional molecule is modeled in two dimensions. A model may give the bonds in a molecule correctly but perhaps not the spatial configuration of the atoms. So one might say that a model "conforms to reality" in certain respects. It follows that there can be models of the same molecule that have different degrees of sophistication. However, while "no model can convey everything," according to Russell, models can still be "right" or "an approximation to the truth, to reality":

> [E]ach time we are getting closer and closer I think to the reality that's actually there.
>
> *Colin Russell, chemist and historian of chemistry*

Thus Russell's realism is tempered in that models are realistic with regard to certain aspects of a phenomenon but not with respect to others.

Some scientists highlight that the link between model and reality is feeble. Solid state physicist John Bolton put this into words when he talked about a toy world created by the model.

> It's not the real world. It's a toy world, but you hope it captures some central aspects of reality, qualitatively at least, how systems respond.
>
> *John Bolton, solid state physicist*

Bolton gives as an example the Ising model for the behavior of physical systems at some critical temperature, a model that acquired a life of its own, for some time at least (cf. Hughes 1999), such that these models are examined for their own sake and not because they are a model of something. Sometimes an application is subsequently found, or even fabricated, for such models. However, ultimately the real world is not given up for the toy world.

> But I think the overwhelming motivation must be that models are meant to be something about reality.
>
> *John Bolton, solid state physicist*

Some of the scientists use the term of *representation* to describe the relationship between models and reality, which is in agreement with some current philosophical literature.

> Any representation of what is going on at the molecular scale by definition must be a model, because it can't be the real thing. Any representation whatsoever, be it of a simple structural formula, a steric formula, a confirmational formula, . . . a formula of showing all kinds of things getting together and rearranging and reacting, all of that, is a model.
>
> *Colin Russell, chemist and historian of chemistry*

Importantly, as Russell outlines, there are different modes in which a model can be a representation. Models as used by the theoretical physicist are certainly at one extreme; the way they represent is by consisting of a set of equations, as Lambourne states:

> So the real world does what it does and the model is some simplified representation of part of it in one form or another, and in my case generally in the form of equations.
>
> *Robert Lambourne, particle physicist*

The biogeochemist Nancy Dise also summarizes:

> That must be what I would consider a model. It's a representation of the system that you study using simple, using the most important, but simple parameters.
>
> *Nancy Dise, biogeochemist*

This brings up the question of what representation means. How "real" are models? Dise suggests:

> It's not real in the sense that it doesn't count everything . . . that's going on. So if you call that not real, then it's not real.
>
> *Nancy Dise, biogeochemist*

And yet, there is the search for those elements that "drive the system." The model about the microbial process "will never be exactly, exactly to the tenth of a degree, what happens," says Dise, "but it's pretty much, you'd say, that's a real thing."

However, representation is not the only concept chosen to describe the relationship between model and reality. There is "mirror" in the case of Malcolm Longair and "correspondence" used by some others, such as Ray Mackintosh. It would nonetheless seem that *representation* is a term frequently resorted to in order to describe the relationship of models to reality. Models are not quite how things are in reality, although they come close, or are intended so. Calling what models do "representation" does not, however, tell us much about what such a representational relationship involves.

8.2 Characterizations of Representation

Having glimpsed the range of possible interpretations of the claim that models represent reality, I now turn to the philosophical treatment of this issue. Considering the relationship of models to the world, a recurrent theme is that models are neither true nor false (for example, Hutten 1954, p. 296). Such "knowledge disclaimers" ("merely hypothetical," "not true," "neither true nor false") are somewhat in contrast to the importance that scientists and philosophers alike attribute to scientific models. Also, a false model need not necessarily be a useless model (Wimsatt 1987). The study of scientific practice continues to reinforce the view that models are crucial elements in developing scientific accounts of empirical phenomena (for example, Cartwright, Shomar, Suárez 1995; Hartmann 1995 and1997; Hughes 1997b; Morrison 1997; contributions to Morrison and Morgan 1999 and Bailer-Jones 2000). This endorses the need to characterize models, well beyond having a purely heuristic role, as indispensable for doing science and therefore as important carriers of scientific knowledge.

Attempts have been made to avoid the inappropriate descriptors of truth and falsity for scientific models because they are seen as restricting the range and possibilities of exploring scientific models unduly. The reason is not that models cannot be true or false per se. The reason is also not that there is no way we could find out whether a model is true or false. The difficulty is rather how to establish whether a model is true or false, given that it may be true in some respects and false in others. How should one classify models, given that most of the time we know that they are partly true and partly false? If the fact that models are often partly false sufficed to discount them, then there would be absolutely no point in paying attention to models. If one wants to study models, then the knowledge that they sometimes make certain false claims must be tolerated. The term of *representation* introduces a relationship between a symbolic construct, a model, and a phenomenon or object of the world, which could, in epistemological terms, be considered a mixture between truth and falsity. The problem is that models do not *mirror* or *replicate* the world in every single detail. Some may object and point out that it is even hard to conceive what "mirroring" or "replicating" should mean.[1]

The terminology of "replication," "copying," or "mirroring" is frequently cause for misunderstanding and miscomprehension. I take them to be used to express the view that somehow all structure and all features of a phenomenon can reappear in the model. Thus there exists the ideal of getting every single detail right. This is, of course, something likely to be impossible. So the relationship that the concept of representation is expected to capture is that symbolic constructs—models—stand in some relevant relationship to the world, without being *replications* of it. They do not "map onto or correspond to the world in a determinate way" (Morrison 1998, p. 66). Incidentally, it is indicative of the difficulty of grasping the concept of representation that what it means for models to represent is, mostly, defined negatively—namely, as *not* mirroring or *not* replicating.

Much of the discussion of how models represent has been conducted drawing from the analogy of representation in art, as, for instance, influentially explored by Nelson Goodman (1976). Goodman turns against what he calls the "copy theory" of representation.[2] He argues that representation has nothing to do with resemblance. One of Goodman's points is that there exists no "innocent eye." We always see something *as* something. He elaborates: "'To make a faithful picture, to come as close as possible to copying the object just as it is.' This simple-minded injunction baffles me; for the object before me is a man, a swarm of atoms, a complex of cells, a fiddler, a friend, a fool, and much more. If none of these constitute the object as it is, what else might? If all are ways the object is, then none is *the* way the object is. I cannot copy all these at once; and the more nearly I succeeded, the less would the result be a realistic picture" (ibid., pp. 6–7). The conclusion is, for art at least, that representation is not imitation. Although we can represent something *as* something, we cannot copy it *as* something. Whenever one copies something, it is aspects of it that one focuses on and imitates because there is no innocent eye. This results in an "inability to specify what is to be copied" (ibid., p. 9).

Goodman also stresses that resemblance is not a sufficient condition for representation because something may resemble itself, but this does not mean it is a representation of itself. Moreover, resemblance is a symmetric relationship, if A resembles B, then B resembles A. Yet if a painting represents the Duke of Wellington, then it is still not the case that the Duke of Wellington represents the painting. Goodman here captures our common intuitions about representation and relies in his analysis on our use of language. A further argument against analyzing representation in terms of resemblance is that something may be represented even though it does not exist. A picture may represent a unicorn, yet "where a representation does not represent anything there can be no question of resemblance to what it represents" (ibid., p. 25). "Representation-as" becomes a matter of exemplification rather than denotation (ibid., p. 66) because there simply is nothing that can be denoted.

It remains to be seen to what extent the analogy to art, contemplating the relationship between a painting and reality, illuminates the relationship between scientific models and reality. Returning to models, why do models not tend to be copies of reality? What is it that gave models the reputation of not being true? Well, models have several perceived shortcomings. Simplification and idealization play a role in their formulation, and multiple models of a single object can exist that make contradicting claims (Hutten 1954, p. 298; and Hartmann 1998). Such tensions or inconsistencies between models of the same object or phenomenon arise when the models rely on different sets of principles that are at odds with each other (Hughes 1997b, pp. 334ff.). This poses an epistemological problem if the aim is to give a full and consistent account of a phenomenon that is valid not merely within a limited domain. (This is the notion of representation-as as an as perfect as possible copy.) When several models of one and the same phenomenon exist, the individual models tend to pick up on a selection of features (aspects of the phenomenon) only and do not address the phenomenon in its entirety. Certainly, when studying scientific practice, models often reveal themselves as being one or more of the following: inaccurate (for example, idealized with regard to the available empirical information), inconsistent (with "neighbouring" models or with generally accepted principles), incomplete (regarding the empirical phenomenon they address) (for examples, see Bailer-Jones 2000a).

In this case, how can models be useful? The shoulder-shrugging answer is, they simply are. Yes, they have all these "unreasonable" properties, and yet the practitioners of science are successful in employing them "to push back the frontiers of knowledge"—which must lead us to conclude that models are important forms of knowledge (or of developing knowledge) after all. In other words, models can carry crucial and illuminating information about empirical phenomena despite potentially being inaccurate, inconsistent, and/or incomplete. This is where the concept of representation enters the debate. However, it is a troubling question to decide what representation means here. Thus the concept of representation is a stepping-stone in explaining how scientific models, with their given properties, can be paramount tools of scientific development and carriers of scientific knowledge. As Friedel Weinert (1999, p. 307) has written: "[A]t least some types of models in science are understood as representation of some physical system in a fully realistic sense."

Wanting to give due credit to the role of models in theoretical scientific practice, it is not enough to characterize them as provisional carriers of knowledge; the way models are used in practice involves making positive statements about how things are in the empirical world, while accepting that not everything in the model is also found in the world, and not everything in the world can be found in the model. This acceptance acknowledges models as *dynamic* forms of knowledge that are subject to further development. To say that models represent is an attempt to cover these epistemologically difficult features of one of the very central tools of scientific practice, but this still falls short of explaining what representation involves.

8.3 Truth of Models?

The baseline of the claim that scientific models represent empirical phenomena is that models are *about* these phenomena. This corresponds to what Goodman (1976, p. 5) has called "denotation" in the context of a picture: "The plain fact is that a picture, to represent an object, must be a symbol for it, stand for it, refer to it; and that no degree of resemblance is sufficient to establish the requisite relationship of reference. Nor is resemblance *necessary* for reference; almost anything may stand for almost anything else. A picture that represents—like a passage that describes—an object refers to and, more particularly, *denotes* it. Denotation is the core of representation and is independent of resemblance."

However, considering the representation of fictional entities, such as unicorns, Goodman does not make denotation a requirement for representation. Denotation may be at the core of representation in many cases, but it is also not enough to define the representational relationship, even in the case where denotation takes place. A scientific model relates to reality in more than merely a denotational sense. The traffic sign that is a triangle, white inside and red around the edge, with one corner pointing down, conventionally conveys the message "Give way." This is what this traffic sign denotes, although there is nothing about this sign specifically, other than convention, that indicates that this is what the sign denotes. The color red is often used for warnings, but otherwise there is nothing about the relationship between the red-and-white triangle and the message "Give way" that would indicate that the sign means what it means. Models, in turn, are expected to do more than denote certain phenomena. They are expected to relate to some of the detailed information we have about a phenomenon modeled. It is also possible to interpret models as being about fictional entities. However, I specified that I consider models that are about phenomena (chapters 1 and 8), so I always take it for granted that these models denote something in the real world—a phenomenon.

Models are, in the widest sense, a description of the phenomena they are about. "Description" is here intended as a term wide enough to admit various different forms of external representations, propositional and nonpropositional—for example, texts, diagrams, plots of empirical data, objects, or mathematical equations. These "external representations" are tools in terms of which the model is expressed. These tools are chosen such that the description is not purely phenomenal, but one that aims at an interpretation of the phenomenon in question (see chapter 1). It is, of course, a major difficulty that "the representations used in physics are not, in any obvious sense, all of one kind" (Hughes 1997a, p. S325). This is why R. I. G. Hughes suspects that "[t]he characteristic—perhaps the only characteristic—that all theoretical models have in common is that they provide representations of parts of the world, or of the world as we describe it" (ibid., p. S325).

This points to an intrinsic problem. It is tempting simply to state that models are representations, representations that do the representing in a whole range of different ways. Yet this constitutes no progress if it is as hard to say what a representation is as what a model is. Not all representations are models (for example, paintings) and not all models represent the world in the same way. Going back to Goodman, models that interpret that which they are a model of are somewhat like pictures that represent something *as* something. Clearly, if a model relates to reality in ways more significant than arbitrarily denoting a phenomenon, then the model can do its job better or worse (whatever this means precisely)—just like some paintings do better at getting across a certain idea than others, no matter how much their painter may have intended to represent such and such. So the hope is that some models are "nearer to the truth" than others and, importantly, that we are able to distinguish between them. If we say that models are neither true nor false, then this would imply that at the very least we would need another set of criteria to inform us about the merits of certain models, in contrast to others.

One way of illustrating the problem of the truth and falsity of models is to think of models as *entailing* propositions. I do not mean logical entailment here, as in first-order propositional or predicate logic, where the truth of certain propositions can be deduced from the truth of certain other propositions. Rather, the issue is that in modeling, some means of expression that are employed to capture the content of a model are not propositional. Besides texts and equations, use is made of diagrams, plots of empirical data or objects, to name but a few. These are convenient, appropriate, and efficient means of conveying information about the model and about the phenomenon (Bailer-Jones 2002b). Because they are the means of expression of choice, it may sometimes be hard to express the same kind of information with different means—for example, with pure text. This is not the point at which to make a judgment about whether "a picture is worth more than 10,000 words" (Larkin and Simon 1987). What matters for talk of truth and falsity of models, however, is that at least some of the content of the model *can* be expressed in terms of propositions about the phenomenon modeled. Consider the example of a model of a water molecule made from rods and balls. One piece of information that can be extracted from that model is the angle that the hydrogen atoms and the oxygen atom enclose. There may be more to this model, but this is certainly one bit of the model that can be expressed in the form of a proposition. One may simply say that the H-O-H angle enclosed by the atoms in a water molecule is 104.45 degrees. This proposition would then be *entailed* by the model (Bailer-Jones 2003).

Thinking about models in this way, the consequence of some of the shortcomings of models is that, on occasion, they entail false propositions. (Incidentally, this may be the root for the intuition that models are "neither true nor false."[3]) Note that my claim is not that models exclusively consist of propositions, true or false. However, they can be viewed as entailing them. So if models entail more than one proposition and some of the propositions entailed by the model are false, then it automatically becomes hard to decide whether the model as a whole is true or false. Let me give a couple of fairly straightforward reasons why it may be hard to decide on the truth or falsity of a model.

- It is not clear how many propositions the model entails. This makes it impossible to seek a specific percentage of true propositions in a model. It would seem that one could increase this number at will (for example, by adding conjunctions of true propositions) and fudge any percentage.
- It may not be possible to convey all the information contained in pictures and diagrams into propositions. So there would be ele-

ments (the "pictorial elements") in models to which truth or falsity cannot be attributed simply because they are not propositional.

- To reiterate, my claim is not that models exclusively consist of propositions, true or false. But they can be viewed as *entailing* them. On the contrary, models often contain nonpropositional components.

Why do models ever entail false propositions? Models as we encounter them in scientific practice have certain features, or if you want, "shortcomings," which result in the falsity of certain propositions entailed by models.

Inaccuracy

Models frequently describe empirical phenomena only inaccurately—that is, they are based on approximations and simplifications. For the truth of a model, it makes a difference whether it is the conceptual representation of a phenomenon that is idealized or the problem situation itself (McMullin 1985).

- If it is the *conceptual representation* that is idealized or simplified, it is perhaps mostly a question of degree whether a proposition entailed by a model can still count as true—for example, when assuming that the Earth orbits around the Sun on a circular path, rather than an elliptical path. Perhaps we are inclined to say that to assume a circular path is false. But the path is also not a perfect ellipse, only in approximation.

- If it is the *problem situation* that is idealized, then the resulting model is more obviously false—for example, if, to model the orbit of the Earth around the Sun, one makes the simplifying (and strictly not correct) assumption that besides the Earth and the Sun, there are no other bodies that exert gravitational forces on the Sun and the Earth.

Inconsistency

Certain models may contradict accepted principles or known facts about phenomena. According to Bohr's model of the atom, for example, electrons are supposed to move on circular orbits that correspond to discrete energy levels—that is, have fixed radii. However, as charged accelerated particles, they should be expected to lose energy through radiation, thus gradually move on smaller and smaller orbits corresponding to lower energy levels. Surely, not both can be true. So is it true that electrons move on fixed orbits around the nucleus?

Incompleteness

Most models focus on aspects of a phenomenon only, not on the phenomenon in its entirety. The wave-particle dualism in quantum mechanics is a famous example of this; another is whether one models water as an ideal fluid (continuous and incompressible, obeying the Bernoulli equation) or as consisting of discrete particles, as in a gas. The former allows to model water flow, the latter to model phenomena such as the diffusion of a drop of ink in water. Fluid and particles are concepts that conceptually do not agree with each other. Each model leaves out important aspects of the phenomenon called water. Different models of one and the same phenomenon rarely complement each other perfectly, which means that individual models are bound to entail false propositions regarding some aspects of a phenomenon.

The fact that perfectly useful, fruitful, and acceptable models can entail false propositions sheds some light on a relationship between models and reality, a relationship that may be perceived as problematic by a strong realist. On the one hand, a model is supposed to tell us what an empirical phenomenon is like, and on the other hand, we find that this model may, in the course of it, tell us some "lies." So how and in what sense does the model represent a phenomenon? The beginning of a solution could lie in knowing *which* propositions entailed by the model are true about the phenomenon modeled, and which are not. In other words, there ought to be a rationale for tolerating certain false propositions in a model. Needless to say, the working assumption of the whole modeling approach is that models must have some merits other than telling the truth about a phenomenon in every single respect, because, as I have sketched, this is not something models necessarily do. How can representation work in such a context? There are some other considerations that enter into models telling us what the empirical world is like. Next I suggest some.

Models practically never aim to describe a phenomenon in its entirety. They set themselves smaller, manageable tasks and focus only on selected aspects of phenomena. This is why it is possible to have various different models of one and the same phenomenon that do not necessarily interfere with each other. These *submodels* may each address a different aspect of the phenomenon. In many cases modeling may become possible only when one focuses on one aspect in isolation and when one disregards others (construct idealization). Also, the task of modeling only an aspect of a phenomenon is bound to be easier and less complex than trying to model the entire phenomenon with all its features and properties (Bailer-Jones 2000). Such a step-by-step strategy allows scientists to set out by modeling those aspects for which they have a modeling idea available and postpone others that seem to evade their efforts and imagination for the time. Selection of aspects for the purposes of modeling is an accepted and well-practiced creative strategy (Bailer-Jones 1999). The assumption that features of a phenomenon that are not currently subject to modeling can be disregarded for the time being goes hand in hand with the assumption that *other* models may later be developed to complement the first model by addressing those features that were omitted in the first model. A well-known example for this kind of modeling are the various models of the atomic nucleus (see Morrison 1998, pp. 74ff.). Of course, the problem with partial models of a phenomenon is often that these different models contradict each other to some extent, as is the case with the models of the atomic nucleus.

The selection taking place in scientific modeling is also nicely illustrated with the analogy to maps, an analogy that Ronald Giere (1999b) likes to employ. Maps aspire to represent only *some* aspects of the world; the empirical information that is taken account of is carefully selected. Maps present some region of the world from a certain perspective and only up to a certain accuracy—for example, neglecting distance information in a subway map. The quality of a map depends on whether it represents well that for which it has been designed. This involves (1) being very accurate in representing some aspects of the empirical world—for example, not to miss any turnings and junctions on a road map, and (2) disregarding other information—for instance, what crops are planted along the roadside. Map makers and map readers have learned to concentrate on those features that the map represents and not to search for other features in it that the map does not aspire to represent.

A central example for selective modeling are the various models of the atomic nucleus (see also Morrison 1998, pp. 74ff.). They show how different aspects of a phenomenon are selected and modeled separately from each other. Each model is formulated on different assumptions, and each accounts for selected observed properties of the atomic nucleus. The overall binding energy of the nucleus is proportional to the mass of the nucleus (in approximation)—like in a liquid drop. The liquid-drop model does not, however, explain the so-called magic numbers. These are proton and/or neutron numbers at which the nucleus is particularly stable. This is where the shell model comes in, which assumes that the nucleons exist in concentric shells that can only be occupied by a finite number of nucleons. When these shells are full, a particularly stable configuration is achieved. The maximum shell occupancy explains the magic numbers (analogous to the electron shells in an atom), but the ordinary shell model does not, in turn, explain the variation with mass of the binding energy. Other nuclear models then go on to explain defects in the shell model and the liquid-drop model, such as their inability to account for the presence of quadrupole moments in nuclei. Importantly, it is not only the case that different models focus on different aspects of the phenomenon; often they are also based on selected and partially separate sets of evidence.[4]

The selections made in the process of modeling phenomena make sure that models entail certain true propositions, but they indicate that there are many more things that could or should be said about the phenomenon. Moreover, the focus on modeling certain aspects of a phenomenon sometimes leads to the acceptance of false propositions that address other aspects that are not the focus of the current model. Evidently, the falsity of those propositions that are not the focus of a selection is perceived as being less important than the truth of certain other propositions that are the focus of the model.

As the map analogy implies, certain false propositions or "misrepresentations" within a model may be accepted for some "higher purpose." Sometimes it may be more important for a model to meet a certain function than to be entirely truthful in every respect. When Bohr's model of the atom predicted the hydrogen spectrum really well, it may have been less important that some contradictions with classical electrodynamics came with the model. Keeping in mind the function of a model will help to decide in which respects the model must be truthful in order to meet its function.

What are possible functions of models? Well, models may be designed to predict something, to explain, or to instruct. Some models, often based on analogy, are a vehicle for being creative, for exploring a new path of thinking. In this function, models can also be guides for further experimentation. Or a model may be good at predicting certain outcomes without explaining much. An artificial neural network that has been trained to classify objects can be a useful model for classifying objects without giving an account explaining why the objects fall into the particular categories. So a model may be good at predicting without specifying and explaining the processes that led to the predicted results. Other models may instruct us how to use a machine, but this may not give us any clue about how the machine works. In turn, the technician who built the machine and knows the model that tells him how the machine works may not be good at using the machine. Yet other models may be good teaching aids to get across some idea of how something works, but they may not be quite correct enough to provide a comprehensive account of the processes going on in a phenomenon.

The function that a model has provides a criterion for weighing up the importance of certain propositions entailed by the model. The truth of those propositions that are crucial for the model to fulfill its function are obviously more important than others. Knowing the intended function of the model provides a rationale for weighing up the importance of propositions. The user of a washing machine may be taught that the machine "knows" how heavily it is loaded with clothes and "decides" accordingly how much water is needed. Similarly, a technically advanced washing machine "knows" how soapy the washing still is and how often it needs to be rinsed. Of course, propositions about the machine "knowing" and "deciding" anything, as they occur in the users' model, are false, especially from the point of view of the technician who has to *make* the machine use the right amount of water by technical means and who is all too aware that the machine does not know or decide anything.

Consequently, the technician's model will entail quite different propositions describing technical details about regulating the amount of water and the number of rinsing cycles. In the users' model, in contrast, the kind of propositions that need to be true are simply those that state what *happens* in certain cases—for example, "putting in too much soap results in a larger number of rinsing cycles," "less water is used if the machine is only half-full," or perhaps "if the machines is too heavily loaded, it will not run." It then does not matter if other propositions, such as those about the machine *deciding* or *knowing*, are false in a certain sense. This illustrates that alternative models can be valid with respect to the function for which they were formulated and that models are rarely expected to meet all functions they could potentially have. So meeting a particular function can override the disadvantage of a model entailing false propositions.

The model user

Considering the role of selection and function for how models represent brings in a third element on which the former two are conditional. It is through the model users that a model can be *intended* to be a model of certain aspects of a phenomenon and to have a certain function.[5] The model users' activities of intending, choosing, and making decisions account for the fact that models can be about phenomena without always being a perfect match, without telling us how things are *exactly* in the real world. Importantly, only the model users can decide which false propositions can be tolerated as entailed by a model, given its function and the aspect of a phenomenon the model is about. Or, in turn, model users decide which propositions ought to be true for a model to be acceptable.

8.4 Do Models Represent in the Way Art Does?

I suspect it is an advantage and a disadvantage at the same time that the concept of representation is frequently studied in the light of analogies. There is Giere's analogy between maps and models, which I have already outlined. There the claim is that models *fit* the empirical world just like geographical maps do (Giere 1999b and also Giere 1999a, pp. 25 and 81–82).[6] The other prominent analogy is between models and the way in which pieces of art, such as paintings, represent reality (Goodman 1976; Suárez 1999 and 2002; and French 2003). The problem with analogies is that, although they enrich our understanding of what representation can be and how the term is used in various contexts, they ultimately do not provide definitive clues about how it is that *scientific models* represent. This is a question that needs to be considered separately, perhaps fruitfully in comparison with maps or paintings.

One message that Hughes takes from Goodman's analysis of representation in art is that representation need neither imply similarity nor resemblance. Maybe so, but how else would representation be established in modeling? Another message is that representation is always representation *as* something. This seems to be self-evidently the case for models. Mauricio Suárez (1999) has argued on the basis of pieces of art that representation does not depend on similarity or preservation of structure. One example he uses is Francis Bacon's paintings, copying a painting of Pope Innocent X by Diego Velázquez. Even though Bacon's paintings may have considerable similarity with the Pope, they are not paintings of the Pope and the intention is not to represent the Pope. Moreover, paintings may be taken to represent something in addition to the objects and people on the painting. Suárez's example is Pablo Picasso's *Guernica* representing the threat posed by fascism in Europe. It seems to me the latter example is best interpreted as a metaphor that works at various levels.

There are, of course, things depicted on the painting—such as a wild horse, the torso of a bull, or a crying mother holding up a baby—expressing the pain of the inhabitants of Guernica, but they also stand for something other or something more than what is depicted. This is perhaps best seen as a case of a painting having the capacities of metaphor, just like "Alice is a bookworm" is not uttered to say that Alice is a little, book-destroying insect. The message is more than "meets the eye" or what a statement "literally" says, and sometimes there may be several messages, some "literal," some "metaphorical," as, for example, in "no man is an island." It seems that models may or may not give rise to such different levels of interpretation. It is clear that they often do not. When they simply describe what constitutes a phenomenon, then there may simply be no other level of interpretation. A model of DNA shows where all the different atoms and molecules go and how they are linked to make up the configuration that is DNA, and not much more.

Suárez employs the analogy between how models represent and how paintings represent in order to highlight how important intention is to representing. It matters whether Bacon painted to depict the Pope or not. The intention of the model users is also likely to be an element in a model representing a phenomenon, but we expect other, more publicly accessible criteria to be available regarding when a model represents a phenomenon.[7] For instance, in DNA the water molecules must be distributed around the double helix in a way that is in agreement with existing chemical know-how. Typically, it is the intention of a whole community, founded on some factual information, that leads to a model being taken to represent a phenomenon. In the case of DNA, the community must have ways of "reading" the model, ways of symbolizing atoms and molecules in the structure efficiently. In the case of scientific models, there is no single "artist" in the center of the activity of representing. Moreover, to argue that paintings do not (necessarily) represent by preserving structure is no proof that models do not do this. It seems to me that this is a question that needs to be considered separately. Moreover, there may be senses in which models share the structure of the phenomenon that they model, but that alone may not be enough to constitute the representational relationship.

Steven French (2003, p. 3) has rightly pointed out that structure does not come into art in the same way as it comes into scientific theorizing. The same theoretical structure may be expressed by different mathematical formalisms, perhaps by different people, as happened with the Schrödinger equation and matrix mechanics in the early days of quantum theory. We may also ask whether a theory or a model is still the same after it underwent some changes, but perhaps preserved some of its structure—where does the old theory end and the new one begin? For instance, is Langrangian or Hamiltonian mechanics still the same theory as Newtonian mechanics? In art some piece of art may influence other pieces of art, but they clearly *are* other pieces of art, whether they display something like a similar "structure" or not. A theory or model is obviously more separate from its creator(s) with his or her intentions than pieces of art. French turns the argument around and examines to what extent art represents like scientific models do. In this context he concludes that one can also argue for the relevance of the preservation of structure in representation in art.

In summary, one may explore the analogy between paintings and models, but cannot infer from the analogy what must hold for representation in the case of models. As is clear from the few examples Suárez discusses, paintings are capable of representing a range of concrete and less concrete things, or of not representing at all,[8] in a multitude of different ways, and these ways are hard to analyze comprehensively, even just for the case of art. However, even if this were done, studying the analogy would not save us the effort of examining separately how it is that scientific models represent.

8.5 Do Models Have the Same Structure As the Phenomena Modeled?

Goodman suggests that the basic relationship referred to as "representation" is that one thing, A, *stands for* another, B, or that A is a symbol for B. Hughes (1997a, p. S330) goes along with this notion of *denotation.* One can, for instance, decide that the coins on the table stand for football players. If A stands for B, then A can be taken (in the mind) to denote B temporarily, especially if the fact that A denotes B is useful to those who deal with B through A. A is taken to stand for B for a specific purpose and with a specific aim in mind. For instance, having coins on a table allows us to view the configuration of the players on the football pitch—that is, certain specific relationships, on a small scale (highlighting where the players are in relation to each other at a certain moment in time). There is a certain amount of structural information involved in representing the football players by coins, which can make it possible to *demonstrate* something about the football players by considering the coins. The final step, after denotation and demonstration, is *interpretation* for the model to yield predictions about the real world. It is these three factors in terms of which Hughes (1997a, p. S329) proposes to study representation. With the help of denotation, it becomes possible to demonstrate something within the model that, appropriately interpreted, illuminates the phenomenon that was so modeled.[9]

Hughes's DDI (denotation, demonstration, interpretation) approach to representation provides a skeleton, and as a skeleton there seems not much wrong with it. The burning question remains, however, how it is that there is something about the model that allows us to demonstrate something that then, after appropriate interpretation, becomes applicable to and insightful about real-world phenomena. There is a strong intuition that there is something that the phenomenon and the model share that allows us to treat them in parallel. This "something" is often said to be structure. Emphasis on structure is an important element in the Semantic View of theories (see chapter 6, section 6.1), but also comes up quite naturally in the context of discussing representation. The structure could be something a model and a phenomenon have in common, and sharing a structure would then explain why inferences can be drawn from models about phenomena. However, there exist a number of powerful arguments against accounting for representation solely in terms of structure.

The strongest conceivable claim would be the following: A model represents a phenomenon if and only if the model and the phenomenon have the same structure—that is, are isomorphic to each other. The problem with defining representation in this way is that some of the logical properties of representation and isomorphism are not identical. The relation of isomorphism is symmetric: if A has the same structure as B, then B has the same structure as A. In contrast, the relation of representation is asymmetric: that A is a representation of B does not mean that B is a representation of A. A model may represent water as H_2O, as hydrogen and oxygen atoms in a certain configuration, but this does not mean that water represents that model. The same problem arises with transitivity. If A has the same structure as B, and C has the same structure as B, then we also know that A has the same structure as C. In modeling, we may be able to represent sound as a wave and light as a wave, but this does not mean that light represents sound or that sound represents light.

Finally, the relationship of isomorphism is reflexive. A will always have the same structure as itself, while one would not say that a model represents itself. All this shows is that having the same structure is insufficient basis for claiming a relationship of representation between two things. It is simply not enough for the coins on the table to have the same spatial distribution as the players on the football ground. This on its own does not suffice as a basis for saying that the coins represent the football players. More criteria must apply. These criteria can be roughly characterized as taking the function of the model into account and specifying the aspect of a phenomenon that the model is about, both of which depend on the intention of the model users (see section 8.3). So although the strengths of a model—the fact that a model works well for what it is intended—have frequently to do with having the same or a similar structure, pointing to structure is not enough to illuminate the relationship of representation.

Another problem with a structural account is that it leaves little scope for misrepresentation. Either a model has the same structure as a phenomenon and therefore is a representation of it, or it is not a representation. If it turns out after some research that a model fails to capture the structure of a phenomenon correctly, then the conclusion would have to be not that the model is a bad representation of the phenomenon (and that the model is a bad model of the phenomenon), but that the model never represented the phenomenon in the first place. This seems counterintuitive because the model, no matter what its failings may turn out to be, had been *designed* to be a model of the phenomenon in question, and may, as such, at least have been of temporary benefit. In practice, situations arise where models are replaced by other, better models. The earlier as well as the later models have certain benefits that grant that these models merit the title "representation of phenomenon x." It would be wrongheaded to claim that the earlier models did not represent certain phenomena, simply because they were superseded later on. After all, the same may happen to currently accepted and taken-to-represent models, given their comparative shortcomings. There needs to be room to decide for one model being better than another for certain purposes, which is also why, with time, one model may be replaced by another more permanently.

Note also that a model is only ever discarded once it is replaced by a better model, not merely because it fails to be identical in structure with the phenomenon modeled. So how can we say that an old model never represented the phenomenon in the first place, given that it may also be the fate of the model currently taken to represent the phenomenon that it, too, will be superseded? I have stressed repeatedly that fully acceptable, working models still have shortcomings. They have their merits, but they are not "perfect." This means that there are better or worse representations, or rather that the model users apply criteria of their choice and convenience to distinguish between models representing phenomena better or worse. This feature of representation cannot be accommodated in an account of representation that solely relies on the preservation of structure in order to constitute the representation relationship.

At least some models capture some of the structure of the phenomena they model, but they do so in a variety of senses.[10] Structure may mean the geometrical configuration, as in a model of a bridge or a toy airplane. Structure may also refer to the type of mathematical equation employed to model a phenomenon. Both light and sound have the "structure" of a wave in that they can be modeled with the help of a wave equation. The diagram of the equilibrium model of stars, discussed in chapter 1, also gives us some aspect of the structure of a star, just not the full story and not with regard to every aspect that could be considered in a star. The multitude of different uses of structure is another reason why reference to structure does not do much work in illuminating the concept of representation. This is a result of the pluralism of ways of doing modeling encountered in chapter 1. Neither models nor representation can be compressed into a uniform and tidy account. As a consequence, while failing to come up with an account of representation, it is still possible to highlight some characteristics of models that represent phenomena.

8.6 Conclusion

In this chapter, I reviewed how models relate to phenomena. Models are said to represent phenomena, which means that they "tell" some things right about phenomena. While scientists view the relationship between models and phenomena in a range of different ways (section 8.1), philosophers try to analyze it more systematically (section 8.2). In particular, I analyzed models in terms of their truth and falsity and the additional constraints needed to do so. The question is under what conditions false propositions entailed by a model can be tolerated (section 8.3). In section 8.4, I argued that the way models represent reality is not the same as the way in which pieces of art represent their subjects. Finally, in section 8.5, I showed that identity of structure is not sufficient to establish a representation relationship.

At the end of this chapter, there does not stand a fully fledged account of representation of scientific models. In fact, doubts arise whether such an account is possible in straightforward terms. At the same time, it is clear that models tell us *some things* about how things really are. Still, it is not much help to summarize what models are by saying that models are representations of phenomena. Although I think this is true, the statement does not provide much philosophical insight as long as we do no better at saying what representation is than at stating what scientific models are. But it has been possible, in this chapter, to study the constraints that apply when models represent. The representational relationship is constituted by model users agreeing on the *function* of a model and on the *aspects* of a phenomenon that are modeled. In addition, the aim is that the model is as closely as possible in agreement with some part of the empirical information available about a phenomenon. In terms of the propositional analysis of models of section 8.3, this means that model users weigh the propositions entailed by a model according to certain criteria (depending on the function of the model and the aspect of a phenomenon modeled) and from this decide which of these propositions are crucial to the acceptance and continued use of the model. Falling short of a full-blown account of representation, highlighting the constraints that apply when models represent may count as an acceptable result, albeit a result as imperfect as models themselves.

Notes

1. However, there is no denying that something like a notion of representation as mirroring exists. Margaret Morrison (1999, p. 60), for instance, has talked about "the traditional notion of representing as mirroring." (Of course, this "traditional notion" may be a straw man to attack—a view that nobody seriously defends—but at least as a straw man it exists.) Morrison suggests that the pendulum model comes close to matching that notion of mirroring. She thinks this because the model can be corrected more and more in order to arrive at a more and more "realistic" picture of a pendulum.

2. When Goodman talks about "copy theory," he does not mean "getting every detail right." He has in mind the resemblance theory, and he seems to equivocate the two expressions, as, for instance, when he says: "Incidentally, the copy theory of representation takes a further beating here; for where a representation does not represent anything there can be no question of resemblance to what it represents" (Goodman 1976, p. 25).

3. Weinert (1999, pp. 307 and 315) has made the important point that the widespread notion that models are false arises because numerous authors only think of analogue models (see chapter 5): molecules *are not* billiard balls in any literal or existential sense. But this is different from characterizing models as hypothetical—that is, from suggesting that *it is not known* whether they are true or false. Moreover, truth and falsity as absolute terms do not accommodate the kind of fine-tuning that allows one to describe models as "partially true" and as getting *some* things right about an empirical phenomenon, but not others, and leave open yet others.

4. For a much more detailed discussion of the physics and of further models of the nucleus, see Eisberg and Resnick 1985, pp. 508–550.

5. Similar conclusions have also been drawn by Andoni Ibarra and Thomas Mormann (1997, pp. 77ff.) and by Bas van Fraassen (2000, pp. 50ff.).

6. Hughes (1997a, p. S330) highlights a negative analogy between maps and *theories*, an analogy originally proposed by Stephen Toulmin (1953, chapter 4). Maps only refer to existing particulars, whereas theories are more general in that they tend to deal with a whole class of systems. (Theories in cosmology are an exception.) So, insofar as models are about particulars, and local rather than global, or at least more so than theories, the analogy between maps and models is somewhat more appropriate.

7. French (2003, p. 13) has commented that "even if it is granted that when it comes to artistic objects, 'almost anything may stand for anything else' (Goodman [1969] 1976, p. 5), it is not clear that this is the case for models. Not anything can serve as a scientific model of a physical system; if the appropriate relationships are not in place between the relevant properties then the 'model' will not be deemed scientific."

8. Suárez (1999) has also compared theories to nonrepresentational paintings, such as those by Piet Mondrian, but whereas theories are often

employed in modeling, Mondrians are not used in Picasso or Velázquez paintings.

9. Ibarra and Mormann (1997) have noted that this process should not be viewed as having only one direction (from phenomenon to model via denotation, then demonstration, and then back from model to phenomenon via interpretation). They suggest that "one can characterize the activity of scientists, be it explanation, or prediction, or conceptual exploration, as an oscillating movement between the area of data and the area of symbolic constructions" (ibid., p. 63). In other words, the aim is not just to choose a symbolic construct that matches the data (that may then lead to further predictions), but the way in which the data are interpreted also depends on the chosen symbolic construct—that is, even denoting contains an element of interpretation, and interpreting constantly feeds back into the choice of denotation (cf. chapter 7).

10. Chris Swoyer (1991) gives an account in which he argues for the importance of structure in representation in the context of surrogative reasoning. Suárez (2002) acknowledges isomorphism as one possible means of representation but denies that it is constitutive of representation. Although he also develops a view of the representational relationship that has surrogative reasoning at its core, the general framework in which he analyzes representation is not too different from mine in spirit.

References

Bailer-Jones, D. M. 1999. Creative strategies employed in modelling: A case study. *Foundations of Science* 4: 375–388.

———. 2000. Modelling extended extragalactic radio sources. *Studies in History and Philosophy of Modern Physics* 31B: 49–74.

———. 2002a. Scientists' thoughts on scientific models. *Perspectives on Science* 10: 275–301.

———. 2002b. Sketches as mental reifications of theoretical scientific treatment. In M. Anderson, B. Meyer, and P. Olivier, eds., *Diagrammatic Representation and Reasoning*. London: Springer-Verlag, pp. 65–83.

———. 2003. When scientific models represent. *International Studies in the Philosophy of Science* 17: 59–74.

Cartwright, N., T. Shomar, and M. Suárez. 1995. The tool box of science: Tools for the building of models with a superconductivity example. In W. E. Herfel, W. Krajewski, I. Niiniluoto, and R. Wojcicki, eds., *Theories and Models in Scientific Processes.* Poznan Studies in the Philosophy of the Sciences and the Humanities. Amsterdam: Rodopi, pp. 137–149.

Eisberg, R., and R. Resnick. 1985. *Quantum Physics of Atoms, Molecules, Solids, Nuclei, and Particles.* 2nd edition. New York: John Wiley & Sons.

French, S. 2003. A model-theoretic account of representation (Or, I don't know much about art . . . but I know it involves isomorphism). *Philosophy of Science* 70: 1472–1483. Supplement: Proceedings of PSA. Available online at http://www.journals.uchicago.edu/toc/phos/2003/70/5.

Giere, R. 1999a. *Science Without Laws.* Chicago: University of Chicago Press.

———. 1999b. Using models to represent reality. In L. Magnani, N. Nersessian, and P. Thagard, eds., *Model-Based Reasoning in Scientific Discovery.* New York: Plenum Publishers, pp. 41–57.

Goodman, N. [1969] 1976. *Languages of Art.* Indianapolis, Indiana: Hackett Publishing Company.

Hartmann, S. 1995. Models as a tool for theory construction: Some strategies of preliminary physics. In W. E. Herfel, W. Krajewski, I. Niiniluoto, and R. Wojcicki, eds., *Theories and Models in Scientific Processes.* Poznan Studies in the Philosophy of the Sciences and the Humanities. Amsterdam: Rodopi, pp. 49–67.

———. 1997. Modelling and the aims of science. In P. Weingartner, G. Schurz, and G. Dorn, eds., *The Role of Pragmatics in Contemporary Philosophy: Contributions of the Austrian Ludwig Wittgenstein Society*, vol. 5. Vienna: Hölder-Pichler-Tempsky, pp. 380–385.

———. 1998. Idealization in quantum field theory. In N. Shanks, ed., *Idealization in Contemporary Physics.* Poznan Studies in the Philosophy of Science and the Humanities. Amsterdam: Rodopi, pp. 99–122.

Hughes, R. I. G. 1997a. Models and representation. In L. Darden, ed., *PSA 1996, Philosophy of Science* 64 (Proceedings): S325–S336. Available online at http://www.journals.uchicago.edu/toc/phos/64/s1.

———. 1997b. Models, the Brownian Motion, and the disunities of physics. In J. Earman and J. D. Norton, eds., *The Cosmos of Science: Essays of Exploration.* Pittsburgh: University of Pittsburgh Press, pp. 325–347.

———. 1999. The Ising model, computer simulation, and universal physics. In Morgan and Morrison, *Models as Mediators,* pp. 97–145.

Hutten, E. H. 1954. The role of models in physics. *British Journal for the Philosophy of Science* 4: 284–301.

Ibarra, A., and T. Mormann. 1997. Theories as representations. In A. Ibarra, ed., *Representations of Scientific Rationality: Contemporary Formal Philosophy of Science in Spain.* Poznan Studies in the Philosophy of the Science and the Humanities. Amsterdam: Rodopi, pp. 59–87.

Larkin, J. H., and H. A. Simon. 1987. Why a diagram is (sometimes) worth ten thousand words. *Cognitive Science* 11: 65–99.

McMullin, E. 1985. Galilean idealization. *Studies in History and Philosophy of Science* 16: 247–273.

Morgan, M., and M. Morrison, eds. 1999. *Models as Mediators.* Cambridge: Cambridge University Press.

Morrison, M. 1997. Physical Models and Biological Contexts. In L. Darden, ed., *PSA 1996, Philosophy of Science* 64 (Proceedings): S315–S324.

Morrison, M. C. 1998. Modelling nature: Between physics and the physical world. *Philosophia Naturalis* 35: 65–85.

———. 1999. Models as autonomous agents. In M. Morgan and M. Morrison, eds., *Models as Mediators.* Cambridge: Cambridge University Press, pp. 38–65.

Suárez, M. 1999. Theories, models, and representations. In L. Magnani, N. Nersessian, and P. Thagard, eds., *Model-Based Reasoning in Scientific Discovery.* New York: Plenum Publishers, pp. 75–83.

———. 2002. An inferential conception of scientific representation. *PSA 2002.*

Swoyer, C. 1991. Structural representation and surrogative reasoning. *Synthese: An International Journal for Epistemology, Methodology, and Philosophy of Science* 87: 449–508.

Toulmin, S. 1953. *The Philosophy of Science: An Introduction.* London: Hutchinson.

van Fraassen, B. 2000. The theory of tragedy and of science: Does nature have narrative structure? In D. Sfendoni-Mentzou, ed., *Aristotle and Contemporary Science.* Vol. 1. New York: Peter Lang, pp. 31–59.

Weinert, F. 1999. Theories, models, and constraints. *Studies in History and Philosophy of Science* 30: 303–333.

Wimsatt, W. 1987. False models as means to truer theories. In M. Nitecki, ed., *Neutral Models in Biology.* Oxford: Oxford University Press, pp. 23–55.

CONCLUSION 9

THIS WORK ON SCIENTIFIC MODELS is now completed. It began in the present day, drawing from interviews with scientists who currently do research and actively employ models. Then I traced the history of the treatment of scientific models from nineteenth-century physics to twentieth-century philosophy of science. Some of this work belongs into the comparatively recent discipline of the history of philosophy of science and as such is a contribution to that discipline. My main gist is nevertheless philosophical. Often, however, being aware of historical developments in philosophy aids our understanding of philosophical issues, past and present. This is especially needed regarding the topic of scientific models. Because this topic, for a long time, has led an existence very much at the margins of mainstream philosophy of science, information about its genesis and even its recent development is not easy to come by. This is certainly my experience, which is why I have written the book I always wanted to have when I first set out to work on models.

One of the major shifts in the consideration of models was from models as physical objects, physically built, to models as symbolical and theoretical constructs. It is only due to this shift that models could become a central topic in philosophy of science. Some of the theoretical constructs that we call models today would have still been called analogies a hundred years ago. Another shift was from disregard of models under the influence of Logical Empiricist philosophy to allowing them space and giving them attention in philosophy of science. This shift has been followed by a further shift, whereby theories are pushed out of the limelight when seeking the central tools for giving scientific accounts of natural phenomena. Models currently attract much philosophical attention and, as I have indicated, rightly so. Of course, this development forces us to rethink the concept of a model, as well as that of a theory. This book is a contribution to this process of rethinking.

Let me recapitulate the findings and the arguments of this book in more detail. In chapter 1, I elaborated what constitutes the core idea of a scientific model: a model is an interpretative description of a phenomenon that facilitates access to that phenomenon. The task in the remainder of the book was to substantiate this approach. I have been able to support this general understanding of models with the results from interviews with scientists. Interviewing scientists is not exactly a traditional technique in philosophy of science, but one warranted by the fact that philosophical claims about science should, to a significant degree, be in accordance with scientific practice—and the practitioners of science are evidently part of that practice. The importance of scientific practice is also a point that I argued in chapter 6 (section 6.2), where I consider case study approaches in philosophy of science.

Philosophical progress builds, a lot of the time, on the disputation of views expressed by others, views that are thought to be mistaken, misguided, or dependent upon a flawed argument. This is the reason why the study of models in science cannot forego the insights, misconceived or not, of former students of models. Understanding such earlier views, however, often requires understanding the more general historical and philosophical background of these views. In some philosophical subjects, these precursor views are well known and well worked out, but this was not so for scientific models. In this long neglected subject, a comprehensive examination of the development of the subject has so far been lacking. When I retraced the use of models in the nineteenth century, to which a number of early twentieth-century philosophers referred, then this was not only done for the historical lesson.

There is also the more general issue of what to make of the shift from physically built models that are mechanical in the most obvious sense to a notion of models as symbolic, much more abstractly represented constructs. I considered recent conceptions of mechanism, in contrast to classical ones, to establish the extent to which these may be (a) continuous with a traditional understanding and (b) still applicable, in an important sense, to modern scientific models. If we can still think of models as mechanistic in some revised sense, then this indicates something about a cognitive preference for mechanistic thinking in modeling phenomena, albeit with an "adjusted" notion of what "mechanism" means. This is an area that most definitely requires further study. My approach in chapter 2 provides historical motivation, but it is still tentative. Research into the role of causality and mechanisms in our thinking, and subsequently for modeling, also takes into account related cognitive science findings in these areas.

Another long-standing proposal is to view model thinking in terms of analogical thinking. Analogy has long been tied to the philosophical study of modeling. As I showed in chapter 3, there can be no doubt that analogies are often instrumental in scientific progress, and much research in cognitive science indicates why. However, contrary to some philosophers' beliefs in the 1960s, I highlight that models are *not* analogies. Models are judged by their own standards—whether they represent (see chapter 8)—and this is so, whether or not analogies aid and contribute to thinking about a model. Research in this area is abundant, although it is not always easy to establish the exact way in which studies in different disciplines are relevant to our understanding on analogy use in scientific modeling.

In chapter 4, I worked through the history of Logical Empiricism and linked it up with the "nonstudy" of models in that period. This was necessary to understand the framework in which philosophers began to argue for the importance of models in science. Some of the literature of the 1950s and on is otherwise hard to understand, with its references to operationalism and the problems or the interpretation of theoretical terms. Only in that context can we appreciate how radical the claim was that models are metaphors. I provided an in-depth and up-to-date analysis of this claim in chapter 5, where I interpret the claim itself as an insightful metaphor from which we can learn a lot. Yet, just like not all models are based on analogies, not all models carry the features of metaphor. So this approach also has its limitations.

Rethinking models also requires rethinking the relationship between models and theories. Although this issue was approached formally with the Semantic View of theories (chapter 6, section 6.1), it has not been properly addressed until recently. Distinctions between models and theories were either vague or variant from author to author. This is a debate that is still going on, and chapter 6 contains my contribution to it. The same goes for the debate about phenomena. The distinction between data and phenomena has only been introduced in the 1980s (see chapter 7). I proposed a correction to the dominant account in that I suggested that phenomena are not fixed. Instead, what we take a phenomenon to be changes and develops in the course of modeling that phenomenon. Correspondingly, phenomenon and model, in cases where a phenomenon is well researched, are closely connected. The ongoing concern with such a proposal is whether it is too antirealist. While I am far from suggesting that phenomena do not exist in the world, many will want to argue that phenomena are more independent of our study of them.

The issue of realism, in the context of scientific modeling, was addressed under the label of "representation." A commonly heard proposal is that models *represent* phenomena. What we take representation to be then has an impact on what we think the relationship between the model and the phenomenon is. In chapter 8, I presented my analysis of representation according to which certain "shortcomings" of models, leading to their occasional, or partial, falsity can be accommodated. This is so even if models are thought to represent phenomena, in the sense of telling us what phenomena are like. So, at the end, there stands my picture of models as representing phenomena by describing them. That models often rely on theories for their formulation ensures that they are linked up with the general framework of the scientific discipline to which they belong.

This book ends with an emphasis on contemporary philosophical issues: how theories and models relate to each other, what phenomena are with respect to models, and what it means for a scientific model to represent a phenomenon. I briefly summarize my main claims:

- Phenomena are facts or events of nature that are subject to investigation.
- Models are interpretative descriptions of phenomena that facilitate access to phenomena.
- It is not theory that tells you what the world is like. Theory is that to which we (sometimes) resort when we try to describe what the world is like by developing models.
- "Abstract," said of theories, means having been stripped of specific properties of concrete phenomena to apply to more and different domains.
- Models, in turn, are about concrete phenomena (or prototypes thereof) that have all the properties that real things have.
- Theories are applied to phenomena only via models—by filling in the properties of concrete phenomena.
- The claim that models *represent* phenomena allows for models to entail *some* false claims with regard to a phenomenon, the purpose nonetheless being that the model also entails many true claims.
- Which false claims can be tolerated in a model depends on the aspect of a phenomenon selected to be modeled and on the function of the model. Correspondingly, a model can only be said to represent a phenomenon if there are model users who make the selection and choose a function of the model.

Some of these claims will doubtless require further examination, some will doubtless be contested, but this also means the debate goes on and models attract further study. This may, in turn, result in additional insight in the issues laid out before the reader. This book is a contribution to making the study of models well informed.

BIBLIOGRAPHY OF CONTEMPORARY WORKS ON SCIENTIFIC MODELS

All references cited in a chapter are listed at the end of each chapter. Here I present a bibliography of works on scientific models, not all of which are explicitly cited in the chapters. It was started with the aim of being comprehensive, and although this now seems impossible to achieve, I have tried to make it as complete as possible. The works are listed by year of publication and sorted alphabetically within each year. For volumes that contain a number of contributions on models, the individual contributions are set off with a small bullet and listed after the citation of the larger work. These individual contributions are in alphabetical order. For monographs that do not deal exclusively with models, I list the relevant chapters and their titles, also set off with small bullets after the citation of the larger work.

Editorial note: Works published since 2003 have been added to the list by Burlton Griffith and Peter Machamer.

1902

Boltzmann, L. [1902] 1911. Model. *Encyclopaedia Britannica,* 11th ed. Vol. 18. Cambridge: Cambridge University Press, pp. 638–640.

1914

Duhem, P. [1914] 1954. *The Aim and Structure of Physical Theory.* Translated from the French 2nd edition. Princeton, New Jersey: Princeton University Press.

- Chapter 4: Abstract theories and mechanical models

1939

Carnap, R. 1939. Foundations of logic and mathematics. *International Encyclopedia of Unified Science.* Chicago: Chicago University Press.

1945

Rosenblueth, A., and N. Wiener. 1945. The role of models in science. *Philosophy of Science* 12: 316–321.

1948

Altschul, E., and E. Biser. 1948. The validity of unique mathematical models in science. *Philosophy of Science* 15: 11–24.

Deutsch, K. W. 1948–1949. Some notes on research on the role of models in the natural and social science. *Synthese: An International Journal for Epistemology, Methodology, and Philosophy of Science* 7: 506–533.

1951

Deutsch, K. W. 1951. Mechanism, organism, and society: Some models in natural and social science. *Philosophy of Science* 18: 230–252.

Meyer, H. 1951. On the heuristic value of scientific models. *Philosophy of Science* 18: 111–123.

1953

Braithwaite, R. [1953] 1968. *Scientific Explanation: A Study of the Function of Theory, Probability, and Law in Science.* Cambridge: Cambridge University Press.

- Chapter 4: Models for scientific theories: Their use and misuse

Gregory, R. L. 1953. On physical model explanations in psychology. *British Journal for the Philosophy of Science* 4: 192–197.

Hesse, M. 1953. Models in physics. *British Journal for the Philosophy of Science* 4: 198–214.

1954

Braithwaite, R. 1954. The nature of theoretical concepts and the role of models in an advanced science. *Revue Internationale de Philosophie* 8 (fasc. 1–2): 34–40.

Hesse, M. B. 1954. *Science and the Human Imagination.* London: SCM Press.

- Chapter 8: Scientific models

Hutten, E. H. 1954. The role of models in physics. *British Journal for the Philosophy of Science* 4: 284–301.

1956

Hutten, E. 1956. *The Language of Modern Physics.* New York: Macmillan.

- Chapter 3: The concepts of classical physics

Sellars, W. 1956. *Empiricism and the Philosophy of Mind.* Minnesota Studies in the Philosophy of Science. Vol. 1. Minneapolis: University of Minnesota Press; reprint, 1997, Cambridge: Harvard University Press.

- Chapter 13: Theories and models

1959

Beckner, M. O. 1959. *The Biological Way of Thought.* New York: Columbia University Press.

- Chapter 3: Models in biological theory

Brodbeck, M. 1959. Models, meaning, and theories. In L. Gross, ed., *Symposium on Sociological Theory.* Evanston, Illinois: Row, Peterson and Company, pp. 373–403.

1960

Harré, R. 1960. Metaphor, model, and mechanism. *Proceedings of the Aristotelian Society* 60: 101–122.

1961

Freudenthal, H., ed. 1961. *The Concept and the Role of the Models in Mathematics and Natural and Social Sciences.* Dordrecht: D. Reidel Publishing Company.

- Apostel, L. 1961. Towards the formal study of models in the non-formal sciences. In Freudenthal, *Concept and the Role of the Models,* pp. 1–37.
- Frey, G. 1961. Symbolische und ikonische Modelle. In Freudenthal, *Concept and the Role of the Models,* pp. 89–97.
- Groenewold, H. J. 1961. The model in physics. In Freudenthal, *Concept and the Role of the Models,* pp. 98–103.
- Kuipers, A. 1961. Models and insight. In Freudenthal, *Concept and the Role of the Models,* pp. 125–132.
- Suppes, P. 1961. A comparison of the meaning and uses of models in mathematics and the empirical sciences. In Freudenthal, *Concept and the Role of the Models,* pp. 163–177.

Harré, R. 1961. *Theories and Things.* London: Sheed and Ward.

- Chapter 3: Models to mechanisms [= Harré 1960]

1962

Black, M. 1962. *Models and Metaphors.* Ithaca, New York: Cornell University Press.

Braithwaite, R. B. 1962. Models in the empirical sciences. In E. Nagel, P. Suppes, and A. Tarski, eds., *Logic, Methodology, and Philosophy of Science.* Stanford, California: Stanford University Press, pp. 224–231.

Suppes, P. 1962. Models of data. In E. Nagel, P. Suppes, and A. Tarski, eds., *Logic, Methodology, and Philosophy of Science.* Stanford, California: Stanford University Press, pp. 252–261.

1963

Hesse, M. 1963. *Models and Analogies in Science.* London: Sheed and Ward. [= Hesse 1966]

Popper, K. R. [1963] 1994. Models, instruments, and truth. In K. R. Popper, *The Myth of Framework: In Defence of Science and Rationality.* Edited by M. A. Notturno. London: Routledge, pp. 154–184.

1964

Achinstein, P. 1964. Models, analogies, and theories. *Philosophy of Science* 31: pp. 328–350.

1965

Achinstein, P. 1965. Theoretical models. *British Journal for the Philosophy of Science* 16: 102–120.

Bertalanffy, L. v. 1965. Zur Geschichte theoretischer Modelle in der Biologie. *Studium Generale* 18: 290–298.

Bombach, G. 1965. Die Modellbildung in der Wirtschaftswissenschaft. *Studium Generale* 18: 339–346.

Franck, U. F. 1965. Modelle zur biologischen Erregung. *Studium Generale* 18: 213–329.

Hartmann, H. 1965. Die spezielle Problematik des Modellbegriffs in der Quantentheorie. *Studium Generale* 18: 259–262.

Hartmann, P. 1965. Modellbildungen in der Sprachwissenschaft. *Studium Generale* 18: 364–379.

Heckmann, O. 1965. Weltmodelle. *Studium Generale* 18: 183–193.

Horner, L. 1965. Modell- und Schemabildung in der organischen Chemie. *Studium Generale* 18: 237–256.

Hund, F. 1965. Denkschemata und Modelle in der Physik. *Studium Generale* 18: 174–183.

Kaplan, R. W. 1965. Modelle der Lebensgrundfunktionen. *Studium Generale* 18: 269–284.

Kaulbach, F. 1965. Schema, Bild, und Modell nach den Voraussetzungen des Kantischen Denkens. *Studium Generale* 18: 464–479.

Kroebel, W. 1965. Nachrichtentechnische Modelle. *Studium Generale* 18: 226–231.

Jammer, M. 1965. Die Entwicklung des Modellbegriffes in den physikalischen Wissenschaften. *Studium Generale* 18: 166–173.

Lüders, G. 1965. Kernmodelle. *Studium Generale* 18: 193–198.

Metzger, W. 1965. Über Modellvorstellungen in der Psychologie. *Studium Generale* 18: 346–352.

Müller, G. H. 1965. Der Modellbegriff in der Mathematik. *Studium Generale* 18: 154–166.

Nagorsen, G., and A. Weiss. 1965. Der Modellbegriff in der anorganischen Chemie. *Studium Generale* 18: 262–268.

Peters, H. M. 1965. Modell-Beispiele aus der Geschichte der Biologie. *Studium Generale* 18: 298–305.

Spector, M. 1965. Models and theories. *British Journal for the Philosophy of Science* 16: 121–142.

Stachowiak, H. 1965. Gedanken zu einer allgemeinen Theorie der Modelle. *Studium Generale* 18: 432–463.

Topitsch, E. 1965. Mythische Modelle in der Erkenntnislehre. *Studium Generale* 18: 400–418.

Verschuer, O. v. 1965. Modelle in der humangenetischen Forschung. *Studium Generale* 18: 334–338.

Vetter, A. 1965. Modell und Symbol in der Strukturpsychologie. *Studium Generale* 18: 352–361.

Weil, K. G. 1965. Modellbildung in der physikalischen Chemie. *Studium Generale* 18: 257–259.

Wendler, G. 1965. Über einige Modelle in der Biologie. *Studium Generale* 18: 284–290.

1966

Ackermann, R. 1966. Confirmatory models of theories. *British Journal for the Philosophy of Science* 16: 312–326.

Hesse, M. 1966. *Models and Analogies in Science.* Notre Dame, Indiana: University of Notre Dame Press. [= Hesse 1963]

Swanson, J. W. 1966. On models. *British Journal for the Philosophy of Science* 17: 297–311.

1967

Farre, G. L. 1967. Remarks on Swanson's theory of models. *British Journal for the Philosophy of Science* 18: 140–147.

Hesse, M. 1967. Models and analogy in science. In P. Edwards, ed., *The Encyclopedia of Philosophy.* New York: Macmillan, pp. 354–359.

Suppes, P. 1967. What is a scientific theory? In S. Morgenbesser, ed., *Philosophy of Science Today.* New York: Basic Books Publishers, pp. 55–67.

1968

Achinstein, P. 1968. *Concepts of Science.* Baltimore, Maryland: Johns Hopkins University Press.
- Chapter 7: Analogies and models
- Chapter 8: On a semantical theory of models

McMullin, E. 1968. What do physical models tell us? In B. van Rootselaar and J. F. Staal, eds., *Logic, Methodology, and Science III.* Amsterdam: North Holland Publishing Company, pp. 385–396.

Mellor, D. H. 1968. Models and analogies in science: Duhem *versus* Campbell? *Isis* 59: 282–290.

1969

Byerly, H. 1969. Model-structures and model-objects. *British Journal for the Philosophy of Science* 20: 135–144.

Fürth, R. 1969. The role of models in theoretical physics. In R. S. Cohen and M. W. Wartofsky, eds., *Boston Studies in the Philosophy of Science,* vol. 5. Dordrecht: D. Reidel Publishing Company.

Schaffner, K. F. 1969. The Watson-Crick model and reductionism. *British Journal for the Philosophy of Science* 20: 325–348.

1970

Harré, R. 1970. *The Principles of Scientific Thinking.* London: Macmillan.

1971

Carloye, J. C. 1971. An interpretation of scientific models involving analogies. *Philosophy of Science* 38: 562–569.

1973

Bunge, M. 1973. *Method, Model, and Matter.* Dordrecht: D. Reidel Publishing Company.

- Chapter 5: Concepts of models

1974

Bushkovitch, A. V. 1974. Models, theories, and Kant. *Philosophy of Science* 41: 86–88.

Leatherdale, W. H. 1974. *The Role of Analogy, Model, and Metaphor in Science.* New York: American Elsevier.

1976

Harré, R. 1976. The constructive role of models. In L. Collins, ed., *The Use of Models in the Social Sciences.* London: Tavistock Publications, pp. 16–43.

McMullin, E. 1976. The fertility of theory and the unit for appraisal in science. In R. S. Cohen, P. K. Feyerabend, and M. W. Wartofsky, eds., *Essays in Memory of Imre Lakatos.* Dordrecht: D. Reidel Publishing Company, pp. 395–432.

1977

Black, M. 1977. More about metaphor. *Dialectica* 31: 43–57.

1980

Beatty, J. 1980. Optimal-design models and the strategy of model building in evolutionary biology. *Philosophy of Science* 47: 532–561.

Leplin, J. 1980. The role of models in theory construction. In T. Nickles, ed., *Scientific Discovery, Logic, and Rationality.* Dordrecht: D. Reidel Publishing Company, pp. 267–283.

Redhead, M. 1980. Models in physics. *British Journal of the Philosophy of Science* 31: 145–163.

van Fraassen, B. 1980. *The Scientific Image.* Oxford: Clarendon Press.

- Chapter 3: To save the phenomena

1982

Cushing, J. 1982. Models and methodologies in current theoretical high-energy physics. *Synthese: An International Journal for Epistemology, Methodology, and Philosophy of Science* 50: 5–101.

1983

Cartwright, N. 1983. *How the Laws of Physics Lie.* Oxford: Clarendon Press.

- Essay 8: The simulacrum account of explanation

Cushing, J. T. 1983. Models, high-energy theoretical physics, and realism. In *PSA 1982*, vol. 2. East Lansing, Michigan: Philosophy of Science Association, pp. 31–56.

1986

Botha, M. E. 1986. Metaphorical models and scientific realism. In *PSA: Proceedings of the Biennial Meeting of the Philosophy of Science Association, 1986.* Vol. 1, *Contributed Papers (1986).* Chicago: University of Chicago Press, on behalf of the Philosophy of Science Association, pp. 159–171.

1987

Wimsatt, W. 1987. False models as means to truer theories. In M. Nitecki, ed., *Neutral Models in Biology.* Oxford: Oxford University Press, pp. 23–55.

1988

Giere, R. 1988. *Explaining Science: A Cognitive Approach.* Chicago: University of Chicago Press.
- Chapter 3: Models and theories
- Chapter 7: Models and experiments

Harré, R. 1988. Where models and analogies really count. *International Studies in the Philosophy of Science* 2: 118–133.

1990

Bernzen, R. 1990. Modell. In H.-J. Sandkühler, ed., *Europäische Enzyklopädie zu Philosophie und Wissenschaften.* Hamburg: Felix Meiner Verlag, pp. 425–432.

Griesemer, J. R. 1990. Modeling in the museum: On the role of remnant models in the work of Joseph Grinnell. *Biology and Philosophy* 5: 3–36.

Morton, A. 1990. Mathematical modelling and contrastive explanation. *Canadian Journal of Philosophy* (suppl. vol.) 16: 251–270.

1991

Cartwright, N. 1991. Fables and models. *Proceedings of the Aristotelian Society* (suppl.) 65: 55–68. [= Chapter 2: Fables and models, in Cartwright 1999a]

Griesemer, J. R. 1991. Material models in biology. In A. Fine, M. Forbes, and L. Wessels, eds., *PSA 1990*, vol. 2. East Lansing, Michigan: Philosophy of Science Association, pp. 79–93.

1992

Downes, S. M. 1992. The importance of models in theorizing: A deflationary semantic view. In David Hull, Micky Forbes, Kathleen Okruhlik, eds., *PSA 1992*, vol. 1. East Lansing, Michigan: Philosophy of Science Association, pp. 142–153.

1993

Black, M. 1993. More about metaphor. In Andrew Ortony, ed., *Metaphor and Thought.* Cambridge: Cambridge University Press, pp. 19–41. [this work is slightly modified from Black 1977]

Hesse, M. B. 1993. Models, metaphor, and truth. In E. J. Reuland and F. R. Ankersmit, eds., *Knowledge and Language. Vol. 3, Knowledge and Metaphor.* Dordrecht: Kluwer Academic Publishers, pp. 49–66. [= Hesse 1995]

Morton, A. 1993. Mathematical models: Questions of trustworthiness. *British Journal for the Philosophy of Science* 44: 659–674.

1994

Horgan, J. 1994. Icon and *Bild*: A note on the analogical structure of models—the role of models in experiment and theory. *British Journal for the Philosophy of Science* 45: 599–604.

1995

Aronson, J. L., R. Harré, and E. C. Way. 1995. *Realism Rescued: How Scientific Progress Is Possible.* Chicago: Open Court.

- Chapter 3: A naturalistic analysis of the use of models in science
- Chapter 4: Some proposals for the formal analysis of the use of models in science
- Chapter 5: The type-hierarchy approach to models

Bhushan, N., and S. Rosenfeld. 1995. Metaphorical models in chemistry. *Journal of Chemical Education* 72: 578–582.

Herfel, W. E., W. Krajewski, I. Niiniluoto, and R. Wojcicki, eds. 1995. *Theories and Models in Scientific Processes.* Poznan Studies in the Philosophy of the Sciences and the Humanities. Amsterdam: Rodopi.

- Cartwright, N., T. Shomar, and M. Suárez. 1995. The tool box of science: Tools for the building of models with a superconductivity example. In Herfel et al., *Theories and Models in Scientific Processes,* pp.137–149.
- Hartmann, S. 1995. Models as a tool for theory construction: Some strategies of preliminary physics. In Herfel et al., *Theories and Models in Scientific Processes,* pp. 49–67.
- Psillos, S. 1995. The cognitive interplay between theories and models: The case of nineteenth-century optics. In Herfel et al., *Theories and Models in Scientific Processes,* pp. 105–133.

Hesse, M. 1995. Models, metaphors, and truth. In Z. Radman, ed., *From a Metaphorical Point of View: A Multidisciplinary Approach to the Cognitive Content of Metaphor.* Berlin: Walter de Gruyter, pp. 351–372. [= Hesse 1993]

Way, E. C. 1995. An artificial intelligence approach to models and metaphor. In Z. Radman, ed., *From a Metaphorical Point of View: A Multidisciplinary Approach to the Cognitive Content of Metaphor.* Berlin: Walter de Gruyter, pp. 165–198.

1996

Giere, R. 1996. Visual models and scientific judgment. In B. S. Baigrie, ed., *Picturing Knowledge: Historical and Philosophical Problems Concerning the Use of Art in Science.* Toronto: Toronto University Press, pp. 269–302.

Hartmann, S. 1996. The world as a process: Simulations in the natural and social sciences. In R. Hegselmann et al., eds., *Modelling and Simulation in the Social Sciences from the Philosophy of Science Point of View.* Theory and Decision Library (series). Dordrecht: Kluwer Academic Publishers, pp. 77–100.

Mayo, D. G. 1996. *Error and the Growth of Experimental Knowledge.* Chicago: University of Chicago Press.

- Chapter 5: Models of experimental inquiry

Mecke, K. R. 1996. Das physikalische Modell—eine quantitative Metapher? In W. Bergem, L. Bluhm, and F. Marx, eds., *Metapher und Modell.* Trier: Wissenschaftlicher Verlag Trier.

1997

Cartwright, N. 1997. Models: The blueprints for laws. In L. Darden, ed., *PSA 1996. Philosophy of Science* 64 (Proceedings): S292-S303.

Falkenburg, B. 1997. Modelle, Korrespondenz, und Vereinheitlichung in der Physik. *Dialektik 1997/1: Modelldenken in den Wissenschaften:* 27–42.

French, S., and J. Ladyman. 1997. Superconductivity and structures: Revisiting the London account. *Studies in History and Philosophy of Modern Physics* 28: 363–393.

Hartmann, S. 1997. Modelling and the aims of science. In P. Weingartner, G. Schurz, and G. Dorn, eds., *The Role of Pragmatics in Contemporary Philosophy: Contributions of the Austrian Ludwig Wittgenstein Society,* vol. 5. Vienna: Hölder-Pichler-Tempsky, pp. 380–385.

Hughes, R. I. G. 1997. Models and Representation. In L. Darden, ed., *PSA 1996, Philosophy of Science* 64 (Proceedings): S325–S336.

———. 1997. Models, the Brownian Motion, and the disunities of physics. In J. Earman and J. D. Norton, eds., *The Cosmos of Science: Essays of Exploration.* Pittsburgh: University of Pittsburgh Press, pp. 325–347.

Hüttemann, A. 1997. Zur Rolle von Modellen in der Physik. *Dialektik 1997/2: Modelldenken in den Wissenschaften*: 141–150.

Manstetten, R., and M. Faber. 1997. Homo oeconomicus: Reichweiten und Grenzen eines Modells. Bemerkungen zu einem Aufsatz von Mary S. Morgan "The Character of 'Rational Economic Man'." *Dialektik 1997/3: Modelldenken in den Wissenschaften*: 121–120.

Morgan, M. 1997. The character of "rational economic man." *Dialektik 1997/1: Modelldenken in den Wissenschaften*: 77–94.

———. 1997. The technology of analogical models: Irving Fisher's Monetary Worlds. In L. Darden, ed., *PSA 1996, Philosophy of Science* 64 (Proceedings): S304–S314.

Morrison, M. 1997. Physical Models and Biological Contexts. In L. Darden, ed., *PSA 1996, Philosophy of Science* 64 (Proceedings): S315-S324.

Morrison, M. C. 1997. Models, pragmatics, and heuristics. *Dialektik 1997/1: Modelldenken in den Wissenschaften*: 13–26.

Wagner, A. 1997. Models in the biological sciences. *Dialektik 1997/1: Modelldenken in den Wissenschaften*: 43–57.

1998

Bradie, M. 1998. Models and metaphors in science: The metaphorical turn. *Protosociology* 12: 305–318.

Cartwright, N. 1998. How theories relate: Takeovers or partnerships? *Philosophia Naturalis* 35: 23–34.

Galison, P. 1998. Feynman's war: Modelling weapons, modelling nature. *Studies in History and Philosophy of Modern Physics* 29: 391–434.

Graßhoff, G. 1998. Modelling the astrophysical object SS433—Methodology of model construction by a research collective. *Philosophia Naturalis* 35: 161–207.

Koperski, J. 1998. Models, confirmation, and chaos. *Philosophy of Science* 65: 624–648.

Morrison, M. C. 1998. Modelling nature: Between physics and the physical world. *Philosophia Naturalis* 35: 65–85.

Muschik, W. 1998. Experiments, models, and theories. Comment on Margaret Morrison. *Philosophia Naturalis* 35: 87–93.

Stöckler, M. 1998. On the unity of physics in a dappled world. Comment on Nancy Cartwright. *Philosophia Naturalis* 35: 35–39.

1999

Bailer-Jones, D. M. 1999. Creative strategies employed in modelling: A case Study. *Foundations of Science* 4: 375–388.

———, and S. Hartmann. 1999. Modell. In H.- J. Sandkühler, ed., *Enzyklopädie der Philosophie.* Hamburg: Felix Meiner Verlag, pp. 854–859.

Cartwright, N. 1999. *The Dappled World: A Study of the Boundaries of Science.* Cambridge: Cambridge University Press.

- Chapter 2: Fables and models [= Cartwright 1991]
- Chapter 3: Nomological machines and the laws they produce
- Chapter 8: How bridge principles set the domain of quantum theory (this is a shortened version of Cartwright 1999b)

French, S. 1999. Models and mathematics in physics: The role of group theory. In J. Butterfield and C. Pagonis, eds., *From Physics to Philosophy.* Cambridge: Cambridge University Press, pp. 187–207.

Giere, R. 1999. *Science Without Laws.* Chicago: University of Chicago Press.

- Chapter 1: Introduction
- Chapter 4: Naturalism and realism
- Chapter 7: Visual models and scientific judgment [= Giere 1996]

Hendry, R. F. 1999. Molecular models and the question of physicalism. *Hyle* 5: 143–160.

Magnani, L., N. Nersessian, and P. Thagard, eds. 1999. *Model-Based Reasoning in Scientific Discovery.* New York: Plenum Publishers.

- Bailer-Jones, D. M. 1999. Tracing the Development of Models in the Philosophy of Science. In Magnani et al., *Model-Based Reasoning*, pp. 23–40.

- Dunbar, K. 1999. How scientists build in vivo science as a window on the scientific mind. In Magnani et al., *Model-Based Reasoning*, pp. 85–99.
- Giere, R. 1999. Using Models to Represent Reality. In Magnani et al., *Model-Based Reasoning*, pp. 41–57.
- Harris, T. 1999. A hierarchy of models and electron microscopy. In Magnani et al., *Model-Based Reasoning*, pp. 139–148.
- Knoespel, K. J. 1999. Models and diagrams within the cognitive field. In Magnani et al., *Model-Based Reasoning*, pp. 59–73.
- Nersessian, N. 1999. Model-based reasoning in conceptual change. In Magnani et al., *Model-Based Reasoning*, pp. 5–22.
- Raisis, V. 1999. Expansion and justification of models: The exemplary case of Galileo Galilei. In Magnani et al., *Model-Based Reasoning*, pp. 149–164.
- Suárez, M. 1999. Theories, Models, and Representations. In Magnani et al., *Model-Based Reasoning*, pp. 75–83.
- Winsberg, E. 1999. The hierarchy of models in simulation. In Magnani et al., *Model-Based Reasoning*, pp. 255–269.
- Zytkow, J. M. 1999. Scientific modeling: A multilevel feedback process. In Magnani et al., *Model-Based Reasoning*, pp. 311–325.

Mainzer, K. 1999. Computational models and virtual reality: New perspectives of research in chemistry. *Hyle* 5: 117–126.

Morgan, M., and M. Morrison, eds. 1999. *Models as Mediators*. Cambridge: Cambridge University Press.

- Boumans, M. 1999. Built-in justification. In Morgan and Morrison, *Models as Mediators*, pp. 66–96.
- Cartwright, N. 1999. Models and the limits of theory: Quantum Hamiltonians and the BCS model of superconductivity. In Morgan and Morrison, *Models as Mediators*, pp. 241–281.
- Hartmann, S. 1999. Models and stories in hadron physics. In Morgan and Morrison, *Models as Mediators*, pp. 326–346.
- Hughes, R. I. G. 1999. The Ising model, computer simulation, and universal physics. In Morgan and Morrison, *Models as Mediators*, pp. 97–145.
- Klein, U. 1999. Techniques of modelling and paper-tools in classical chemistry. In Morgan and Morrison, *Models as Mediators*, pp. 146–167.
- Morgan, M. S. 1999. Learning from models. In Morgan and Morrison, *Models as Mediators*, pp. 347–388.
- Morrison, M. C. 1999. Models as autonomous agents. In Morgan and Morrison, *Models as Mediators*, pp. 38–65.
- Morrison, M. C., and M. S. Morgan. 1999. Models as mediating instruments. In Morgan and Morrison, *Models as Mediators*, pp. 10–37.

- Reuten, G. 1999. Knife-edge caricature modelling: The case of Marx's Reproduction Schema. In Morgan and Morrison, *Models as Mediators*, pp. 197–240,
- Suárez, M. 1999. The role of models in the application of scientific theories: Epistemological implications. In Morgan and Morrison, *Models as Mediators*, pp. 168–196.
- van den Bogaard, A. 1999. Past measurements and future prediction. In Morgan and Morrison, *Models as Mediators*, pp. 282–325.

Salmon, M. 1999. Models for explaining archaeological phenomena. In A. Baccari, ed., *Il Ruolo del Modello nella Scienza a nel Sapere*. Rome: Accademia Nazionale dei Lincei, pp. 111–125.

Tomasi, J. 1999. Towards "chemical congruence" of the models in theoretical chemistry. *Hyle* 5: 79–115.

Trindle, C. 1999. Entering modeling space: An apprenticeship in molecular modeling. *Hyle* 5: 127–142.

Weinert, F. 1999. Theories, models, and constraints. *Studies in History and Philosophy of Science* 30: 303–333.

Winsberg, E. 1999. Sanctioning models: The epistemology of simulation. *Science in Context* 12: 275–292.

2000

Ankeny, R. A. 2000. Fashioning descriptive models in biology: Of worms and wiring diagrams. *PSA 1998, Philosophy of Science Supplement* 67: S260–S272.

Bailer-Jones, D. M. 2000. Modelling extended extragalactic radio sources. *Studies in History and Philosophy of Modern Physics* 31B: 49–74.

———. 2000. Scientific models as metaphors. In F. Hallyn, ed., *Metaphor and Analogy in the Sciences*. Dordrecht: Kluwer Academic Publishers, pp. 181–198.

da Costa, N., and S. French. 2000. Models, theories, and structures: Thirty years on. In D. Howard, ed., *PSA 1998, Philosophy of Science Supplement* 67: S116–S127.

Del Re, G. 2000. Models and analogies in science. *Hyle* 6: 5–15.

Edmonds, B. 2000. Complexity and scientific modelling. *Foundations of Science* 5: 379–390.

Harré, R., J. L. Aronson, and E. C. Way. 2000. Apparatus as models of nature. In F. Hallyn, ed., *Metaphor and Analogy in the Sciences*. Dordrecht: Kluwer Academic Publishers, pp. 1–16.

Hesse, M. 2000. Models and analogies. In W. H. Newton-Smith, ed., *A Companion to the Philosophy of Science*. Oxford: Blackwell Academic Publishers, pp. 299–307.

Keller, H. F. 2000. Models of and models for: Theory and practice in contemporary biology. *Philosophy of Science* (suppl.) 67: S72–S86.

Lazlo, P. 2000. Playing with molecular models. *Hyle* 6: 85–97.

Ramberg, P. J. 2000. Pragmatism, belief, and reduction: Stereoformulas and atomic models in early stereochemistry. *Hyle* 6: 35–61.

Stöckler, M. 2000. On modeling and simulations as instruments for the study of complex systems. In M. Carrier, G. J. Massey, and L. Ruetsche, eds., *Science at Century's End.* Pittsburgh: University of Pittsburgh Press.

Suppe, F. 2000. Understanding scientific theories: An assessment of developments. In D. Howard, ed., *PSA 1998, Philosophy of Science Supplement* 67, pp. S102–S115.

Weinert, F. 2000. The construction of atom models: Eliminative inductivism and its relation to falsificationism. *Foundations of Science* 5: 491–531.

Zeidler, P. 2000. The epistemological status of theoretical models of molecular structure. *Hyle* 6: 17–34.

2001

Ankeny, R. A. 2001. Model organisms as models: Understanding the "Lingua Franca" of the human genome project. In J. Barrett and J. McKenzie Alexander, eds., *PSA 2000, Part I, Contributed Papers, Philosophy of Science Supplement* 68, pp. S251–S261.

Chakravartty, A. 2001. The semantic or model-theoretic view of theories and scientific realism. *Synthese: An International Journal for Epistemology, Methodology, and Philosophy of Science* 127: 325–345.

Hartmann, S. 2001. Effective field theories, reductionism, and scientific explanation. *Studies in History and Philosophy of Modern Physics* 32B: 267–304.

Morton, A., and M. Suárez. 2001. Kinds of models, kinds of validations. In M. Anderson and P. Bates, eds., *Model Validation: Perspectives in Hydrological Science.* New York: John Wiley, pp. 11–21.

Plutynski, A. 2001. Modeling evolution in theory and practice. In J. Barrett and J. McKenzie Alexander, eds., *PSA 2000, Part I, Contributed Papers, Philosophy of Science Supplement* 68, pp. S225–S236.

Smith, S. 2001. Models and the unity of classical physics: Nancy Cartwright's "Dappled World." *Philosophy of Science* 68: 456–475.

Teller, P. 2001. Twilight of the perfect model model. *Erkenntnis* 55: 393–415.

Winsberg, E. 2001. Simulations, models, and theories: Complex physical systems and their representations. In J. Barrett and J. McKenzie Alexander, eds., *PSA 2000, Part I, Contributed Papers, Philosophy of Science Supplement* 68, pp. S442–S454.

2002

Bailer-Jones, D. M. 2002. Models, metaphors, and analogies, In P. Machamer and M. Silberstein, eds., *Blackwell Guide to Philosophy of Science.* Oxford: Blackwell, pp. 108–127.

———. 2002. Scientists' thoughts on scientific models. *Perspectives on Science* 10: 275–301.

Batterman, R. W. 2002. Asymptotics and the role of minimal models. *British Journal for the Philosophy of Science* 53: 21–38.

Harper, L., and W. C. Myrvold. 2002. Model selection, simplicity, and scientific inference. In J. A. Barrett and J. M. Alexander, eds., *PSA 2000, Philosophy of Science Suppl.* 69, pp. S135–S149.

Keller, E. F. 2002. *Making Sense of Life: Explaining Biological Development with Models, Metaphors, and Machines.* Cambridge: Harvard University Press.

- Part 1: Models: Explaining development without the help of genes
- Part 2: Metaphors: Genes and developmental narratives

Lauth, B. 2002. Transtheoretical structures and deterministic models. *Synthese: An International Journal for Epistemology, Methodology, and Philosophy of Science* 130 (1): 163–172.

Magnani, L., and N. J. Nersessian, eds. 2002. *Model-Based Reasoning: Science, Technology, Values.* New York: Kluwer Academic Publishers.

- Bailer-Jones, D. M., and C. A. L. Bailer-Jones. 2002. Modelling data: Analogies in neural networks, simulated annealing and genetic algorithms. In Magnani and Nersessian, *Model-Based Reasoning*, pp. 147–165.
- Boumans, M. 2002. Calibrations of models in experiments. In Magnani and Nersessian, *Model-Based Reasoning*, pp. 75–93.
- Giere, R. N. 2002. Models as parts of distributed cognitive systems. In Magnani and Nersessian, *Model-Based Reasoning*, pp. 227–241.
- Guala, F. 2002. Models, simulations, and experiments. In Magnani and Nersessian, *Model-Based Reasoning*, pp. 59–74.
- Johnson, M. 2002. Metaphor-based values in scientific models. In Magnani and Nersessian, *Model-Based Reasoning*, pp. 1–19.
- Kurz-Milcke, E. 2002. Modeling practices and "tradition." In Magnani and Nersessian, *Model-Based Reasoning*, pp. 127–146.
- Morgan, M. S. 2002. Model experiments and models in experiments. In Magnani and Nersessian, *Model-Based Reasoning*, pp. 41–58.
- Phillips, J., G. Livingston, and B. Buchanan. 2002. Towards a computational model of hypothesis formation and model building in science. In Magnani and Nersessian, *Model-Based Reasoning*, pp. 209–225.

Morrison, M. 2002. Modelling populations: Pearson and Fisher on Mendelism and biometry. *British Journal for the Philosophy of Science* 53: 39–68.

Sterrett, S. 2002. Physical models and fundamental laws: Using one piece of the world to tell about another. *Mind & Society* 5: 51–66.

Wimsatt, W. 2002. Using false models to elaborate constraints on processes: Blending inheritance in organic and cultural evolution. In J. A. Barrett and J. M. Alexander, eds., *PSA2000, Philosophy of Science Suppl.* 69: S12-S24.

2003

Bailer-Jones, D. M. 2003. Realist-Sein mit Blick auf naturwissenschaftliche Modelle. In C. Halbig and C. Suhm, eds., *Was ist wirklich? Neuere Beiträge zu philosophischen Realismusdebatten.* Frankfurt/Main: Ontos-Verlag, pp. 139–161.

Bailer-Jones, D. M. 2003. When scientific models represent. *International Studies in the Philosophy of Science* 17: 59–74.

Da Costa, N. C. A., and S. French. 2003. *Science and Partial Truth: A Unitary Approach to Models and Scientific Reasoning.* Oxford: Oxford University Press.

Fox Keller, E. 2003. Models, simulation, and "computer experiments." In H. Radder, ed., *The Philosophy of Scientific Experimentation.* Pittsburgh: University of Pittsburgh Press, pp. 198–215.

French, S. 2003. A model-theoretic account of representation (Or, I don't know much about art . . . but I know it involves isomorphism). *Philosophy of Science* 70: 1472–1483. Supplement: Proceedings of PSA.

Knuuttila, T., and A. Voutilainen. 2003. A parser as an epistemic artifact: A material view on models. *Philosophy of Science* 70 (5): 1484–1495.

2004

Abrantes, P. C. C. 2004. Models and the dynamics of theories. *Philosophos: Revista de Filosofia* 9 (2): 225–269.

Chen, Ruey-lin. 2004. Testing through realizable models. *Philosophical Review* (Taiwan) 27: 67–113.

Humphreys, P. 2004. Scientific knowledge. In I. Niiniluoto, ed., *Handbook of Epistemology.* Dordrecht: Kluwer Academic Publishing, pp. 549–569.

2005

Bailer-Jones, D. M. 2005. The difference between models and theories. In C. Nimitz and A. Bechermann, eds. *Philosophie und/als Wissenschaft. Hauptvorträge und Kolloquiumsbeiträge zu GAP.5, Fünfter Internationaler Kongress der Gesellschaft für Analytische Philosophie.* Bielefeld, September 22–26, 2003. Paderborn: Mentis Verlag, pp. 339–353.

———. 2005. Mechanisms past and present. *Philosophia Naturalis,* 42 (1): 1–14.

———. 2005. Models, theories, and phenomena. In Petr Hájek, Luis Valdés-Villanueva, and Dag Westerståhl, eds. *Logic, Methodology, and Philosophy of Science: Proceedings of the Twelfth International Congress* [2003]. London: King's College Publications, pp. 243–255.

Henrique de Araújo Dutra, L. 2005. Os modelos e a pragmática da investigação. *Scientiae Studia: Revista Latino-Americana de Filosofia e História da Ciência* 3 (2): 205–232.

Portides, D. P. 2005. A theory of scientific model construction: "The conceptual process of abstraction and concretisation." *Foundations of Science* 10 (1): 67–88.

Rueger, A. 2005. Perspectival models and theory unification. *British Journal for the Philosophy of Science* 56 (3): 579–594.

2006

Boon, M. 2006. How science is applied in technology. *International Studies in the Philosophy of Science* 20 (1): 27–47.

Contessa, G. 2006. Scientific models, partial structures, and the new received view of theories. *Studies in History and Philosophy of Science* 37 (2): 370–377.

Craver, C. F. 2006. When mechanistic models explain. *Synthese: An International Journal for Epistemology, Methodology, and Philosophy of Science* 153 (3): 355–376.

Fisher, G., 2006. The autonomy of models and explanation: Anomalous molecular rearrangements in early twentieth-century physical organic chemistry. *Studies in History and Philosophy of Science* 37 (4): 562–584.

Forster, M. R. 2006. Counterexamples to a likelihood theory of evidence. *Minds and Machines: Journal for Artificial Intelligence, Philosophy, and Cognitive Science* 16 (3): 319–338.

Godfrey-Smith, P. 2006. The strategy of model-based science. *Biology and Philosophy* 21 (5): 725–740.

Gooding, D. 2006. Visual cognition: Where cognition and culture meet. *Philosophy of Science* 73 (5): 688–698.

Griesmaier, F.-P. 2006. Causality, explanatoriness, and understanding as modeling. *Journal for General Philosophy of Science* 37 (1): 41–59.

Hughes, R. I. G. 2006. Theoretical practice: The Bohm-Pines quartet. *Perspectives on Science: Historical, Philosophical, Social* 14 (4): 457–524.

Justus, J. 2006. Loop analysis and qualitative modeling: Limitations and merits. *Biology and Philosophy* 21 (5): 647–666.

Morrison, M. 2006. Emergence, reduction, and theoretical principles: Rethinking fundamentalism. *Philosophy of Science* 73 (5): 876–887.

Moulines, C. Ulises. 2006. Ontology, reduction, emergence: A general frame. *Synthese: An International Journal for Epistemology, Methodology, and Philosophy of Science* 151 (3): 313–323.

Nersessian, N. J. 2006. Model-based reasoning in distributed cognitive systems. *Philosophy of Science* 73 (5): 699–709.

Odenbaugh, J. 2006. Message in the bottle: The constraints of experimentation on model building. *Philosophy of Science* 73 (5): 720–729.

Plutynski, A. 2006. Strategies of model building in population genetics. *Philosophy of Science* 73 (5): 755–764.

Rivadulla, A. 2006. Metáforas y modelos en ciencia y filosofía. *Revista de Filosofía* (Spain) 31 (2): 189–202.

Schaffner, K. F. 2006. Reduction: The Cheshire cat problem and a return to roots. *Synthese: An International Journal for Epistemology, Methodology, and Philosophy of Science* 151 (3): 377–402.

Stemwedel, J. D. 2006. Getting more with less: Experimental constraints and stringent tests of model mechanisms of chemical oscillators. *Philosophy of Science* 73 (5): 743–754.

Thomson-Jones, M. 2006. Models and the Semantic View. *Philosophy of Science* 73 (5): 524–535.

Weisberg, M. 2006. Forty years of "The Strategy": Levins on model building and idealization. *Biology and Philosophy* 21(5): 623–645.

Winsberg, E. 2006. Handshaking your way to the top: Simulation at the nanoscale. *Philosophy of Science* 73 (5): 582–594.

———. 2006. Models of success versus the success of models: Reliability without truth. *Synthese: An International Journal for Epistemology, Methodology, and Philosophy of Science* 152 (1): 1–19.

Winther, R. G. 2006. On the dangers of making scientific models ontologically independent: Taking Richard Levins' warnings seriously. *Biology and Philosophy* 21 (5): 703–724.

2007

Andò, V., and G. Nicolaci, eds. 2007. Processo alla prova: Modelli e pratiche di verifica dei saperi. Roma: Carocci Editore.

Bailer-Jones, D. M. 2007. Operationalism, Logical Empiricism, and the murkiness of models. *Revista Portuguesa de Filosofia* 63 (1–3): 145–167.

Batitsky, V., and Z. Domotor. 2007. When good theories make bad predictions, part I. *Synthese: An International Journal for Epistemology, Methodology, and Philosophy of Science* 157 (1): 79–103.

Bays, T. 2007. More on Putnam's models: A reply to Bellotti. *Erkenntnis: An International Journal of Analytic Philosophy* 67 (1): 119–135.

Contessa, G. 2007. Scientific representation, interpretation, and surrogative reasoning. *Philosophy of Science* 74 (1): 48–68.

Darden, L. 2007. Mechanisms and models. In D. L. Hull and M. Ruse, eds., *The Cambridge Companion to the Philosophy of Biology.* Cambridge: Cambridge University Press, pp. 139–159.

Elgin, C. 2007. Understanding and the facts. *Philosophical Studies: An International Journal for Philosophy in the Analytic Tradition* 132 (1): 33–42 (part 1).

Emch, G. G. 2007. Models and the dynamics of theory-building in physics: Part I—Modeling strategies. *Studies in History and Philosophy of Modern Physics* 38 (3): 558–585.

Krohs, U., and W. Callebaut. 2007. Data without models merging with models without data. In F. C. Boogerd, F. J. Bruggeman, J.-H. S. Hofmeyr, and H. V. Westerhoff, eds., *Systems Biology: Philosophical Foundations.* Amsterdam: Elsevier, pp. 181–213.

Landry, E. 2007. Shared structure need not be shared set-structure. *Synthese: An International Journal for Epistemology, Methodology, and Philosophy of Science,* 158 (1): 1–17.

Lenhard, J. 2007. Computer simulation: The cooperation between experimenting and modeling. *Philosophy of Science* 74 (2): 176–194.

Magnani, L., and Li Ping. 2007. *Model-Based Reasoning in Science, Technology, and Medicine: Proceedings of the International Conference MBR06 China*. New York: Springer.

Martínez, C., J. L. Falguera, and J. M. Sagüillo, eds. 2007. *Current Topics in Logic and Analytic Philosophy/Temas actuales de Lógica y Filosofía Analítica*. Cursos e congresos da universidade de Santiago de compostela, 167. Santiago de Compostela: Universidade de Santiago de Compostela.

- Lorenzano, P. 2007. Exemplars, models, and laws in classical genetics. In Martínez, Falguera, and Sagüillo, *Current Topics in Logic and Analytic Philosophy*, pp. 89–102.

Morrison, M. 2007. Where have all the theories gone? *Philosophy of Science* 74 (2): 195–228.

Politzer, G. 2007. Reasoning with conditionals. *Topoi: An International Review of Philosophy* 26 (1): 79–95.

Rakover, S. S. 2007. *To Understand a Cat: Methodology and Philosophy*. Amsterdam: J Benjamins.

Ramsey, J. L. 2007. Calibrating and constructing models of protein folding. *Synthese: An International Journal for Epistemology, Methodology and Philosophy of Science* 155 (3): 307–320.

Vergauwen, R., and A. Zambak. 2007. A new approach to models and simulations in artificial intelligence. *Logique et Analyse* 50 (198): 179–206.

Weisberg, M. 2007. Who is a modeler? *British Journal for the Philosophy of Science* 58 (2): 207–233.

2008

Hartmann, S., C. Hoefer, and L. Bovens, eds. 2008. *Nancy Cartwright's Philosophy of Science*. London: Routledge.

- Bailer-Jones, D. M. 2008. Standing up against tradition: Models and theories in Nancy Cartwright's philosophy of science. In Hartmann, Hoefer, and Bovens, *Nancy Cartwright's Philosophy of Science*.
- Giere, R. 2008. Models, metaphysics, and methodology. In Hartmann, Hoefer, and Bovens, *Nancy Cartwright's Philosophy of Science*.
- Morrison, M. 2008. Models as representational structures. In Hartmann, Hoefer, and Bovens, *Nancy Cartwright's Philosophy of Science*.
- Nordmann, A. 2008. The reader as model, the model as reader: Hermeneutic moments in Nancy Cartwright's philosophy of science. In Hartmann, Hoefer, and Bovens, *Nancy Cartwright's Philosophy of Science*.

Suárez, M., and N. Cartwright. 2008. Theories: Tools versus models. *Studies in History and Philosophy of Modern Physics* 39 (1): 62–81.

2009

Bailer-Jones, D. M., and C. Friebe. 2009. *Thomas Kuhn*. Paderborn: Mentis Verlag.

INDEX